Lecture Notes in Economics and Mathematical Systems

546

Daniel Quadt

Lot-Sizing and Scheduling for Flexible Flow Lines

 Springer

Author

Dr. Daniel Quadt
Faculty of Business Administration and Economics
Catholic University of Eichstätt-Ingolstadt
Auf der Schanz 49
85049 Ingolstadt
Germany

Library of Congress Control Number: 2004109270

ISSN 0075-8442
ISBN 978-3-540-22325-2   ISBN 978-3-642-17101-7 (eBook)
DOI 10.1007/978-3-642-17101-7

springeronline.com

© Springer-Verlag Berlin Heidelberg 2004
Originally published by Springer-Verlag Berlin Heidelberg New York in 2004

Typesetting: Camera ready by author
Cover design: *Erich Kirchner*, Heidelberg

Printed on acid-free paper   42/3130Di   5 4 3 2 1 0

To whom it may concern

Preface

*Das schönste Glück des denkenden Menschen ist, das
Erforschliche erforscht zu haben und das Unerforschliche
ruhig zu verehren*

JOHANN WOLFGANG VON GOETHE

In the sense of Goethe, I hope this book helps to shift the fine line between the inaccessible and the explorable a little bit in the direction of the latter. It has been accepted as a doctoral dissertation at the Faculty of Business Administration and Economics of the Catholic University of Eichstätt-Ingolstadt, Germany, and has been a great pleasure to write—at least most of the time. At the other times, the following people helped me to get *around* the mountains that seemed too high to climb:

First of all, these were my parents, whom I thank for their continuing support. This 'thank you' is extended to my brothers Marcel and Dominik and of course to Angelika, who has strengthened me with love through the sometimes arduous campaign. Further, I thank my academic advisor Prof. Dr. Heinrich Kuhn and my colleagues Dr. Georg N. Krieg and Florian Defregger for many fruitful discussions and helpful thoughts. I also thank Prof. Dr. Klaus D. Wilde for refereeing the thesis.

Many people have contributed indirectly to this book, mostly by simply making my life worth living: In this sense, a big thank you to all@winfos.com and friends, the Gospel Disco, the PennState class of '98, the 'Kölner Karneval', and alle Vogelfreien dieser Welt. Further, I thank my 'new' friends from Ingolstadt, wherever their life has led them in the meantime. I also thank Alex for some guiding light and David, Eric, Fran, John and Rob from the 'Hooters' for some great music and even better concerts! Finally, a credit to Pat Lawlor and everybody else involved in designing the pinball 'Twilight Zone'. The quotes throughout this book are taken from this pinball.

I wish the reader an exciting and interesting time with this book—don't be scared by science!

Ingolstadt, May 2004   *Daniel Quadt*

This preface contains quotes by the Hooters, the Red Rockers, Soulsister, Bläck Fööss and John Watts/Fischer-Z.

Contents

List of Figures

List of Tables

1

Introduction

You unlock this door with the key of imagination!

1.1 Lot-Sizing and Scheduling

1.2 Flexible Flow Lines

1.3 Characteristics of the Problem

1.4 Scope and Outline of the Study

The recent trends towards a globalized economy open new markets for almost every industry. At the same time, companies are being challenged by competitors entering their home markets, resulting in a strong competition. Many companies are no longer able or willing to face this competition by reducing the price of their products. Instead, they try to increase the quality of the products and services. One way to do this is to better fulfill a customer's needs. Since different customers have different needs, the trend goes from a standardized product to customized variants. Thus, the number of products that have to be produced increases substantially. For example, Thonemann and Bradley (2002) report that in the packaged-goods industry, the number of new products doubled from 12 000 to 24 000 between 1986 and 1996. They also state that in the 1950s, a large supermarket offered a magnitude of 1 000 products, while a modern supermarket in the early 21st century carries 30 000 products. Most other industries, such as the automotive, computer hardware, software and telecommunication industry underwent similar developments.

At the same time, companies also try to reduce the delivery time of their products to the customer.

In order to cope with an increasing number of products and a short delivery time, elaborate production planning systems are needed. Roughly speaking, a production planning system determines which product to produce on which machine at what time. Without these systems, companies cannot optimally deploy their capacities and will ultimately lose money.

Production planning problems are too complex to be solved in a monolithic way. Therefore, production planning systems usually distinguish long-term, medium-term and short-term planning. Obviously, these phases have to be connected in some way in order to deliver viable plans. Several hierarchical planning approaches have been suggested in the literature (see e.g. Hax and Candea 1984, Drexl et al. 1994, or Heizer and Render 2004 for an overview). In hierarchical approaches, upper phases have a general, aggregated view on the overall planning problem, while lower phases become more and more detailed. An upper planning phase takes the specifics of lower phases into account and a lower phase uses the result of an upper phase as its input.

The planning phases can be differentiated as follows: In general, long-term production planning levels the seasonal demand effects on an aggregate basis. It uses long- and medium-term demand forecasts and coordinates them with supply, production and staff deployment. Capacities are usually aggregated on a factory level and products on product type level. A typical time frame is one to several years, with a period length of one to three months.

Medium-term planning is more detailed. Inputs are short-term forecasts and given customer orders. Medium-term planning usually covers the main end-products. Capacities are considered as groups of similar production facilities. A typical time frame is one to several months, for example divided into weekly periods.

The focal point of this study is short-term production planning. It consists of lot-sizing and scheduling.

1.1 Lot-Sizing and Scheduling

A lot-size can be defined as the number of product units produced at once. The pioneering work by Harris (1913) thus defines the problem of lot-sizing in its title: 'How many parts to make at once'. Usually, demand volumes of several periods are combined and in one period. This is useful because setting up a machine to produce a certain product may consume time and money. On the other hand, producing a product in a period other than its demand period incurs either inventory holding or back-order costs—depending on whether it is produced before or after its demand period. Hence, the objective of a lot-sizing procedure is to find optimal production quantities that balance the trade-off between setup, inventory holding, and back-order costs. Lot-sizing procedures tend to have a scope of one to several weeks.

Scheduling is concerned with the detailed problem of when and on which machine to produce a product unit (job). It determines the precise production start- and end-times for all jobs and thus the job sequence on each machine. Usually, scheduling procedures cover a single period of the lot-sizing problem, generally between one shift and a few weeks. Their objectives are time-oriented, such as minimizing the completion time of the last job (makespan), due date fulfillment or minimizing the mean flow time through a system of multiple production stages (time between entering the first stage and leaving the last). After the scheduling phase, the final plan—consisting of exact time and machine assignments for all jobs and stages—is handed over to the shop floor and executed.

Lot-sizing and scheduling play an important role even if setup, inventory holding and back-order costs are negligible. In this case, it is still important to create a feasible plan that takes the scarce machine capacity into account (see Tempelmeier 2003, pp. 350–351). Otherwise, the capacities will be left unused or long delivery times will occur.

If the jobs to be scheduled are similar—for example when several units of the same product have to be produced—a batching procedure is usually embedded in the scheduling phase. Batching means to determine how many units of a product to consecutively produce on a specific machine. When setups have to be performed, successively producing more than one unit of the same product reduces setup times and/or costs. Hence, the batching procedure implicitly determines the number of setups per product. A summary on research that integrates batching decisions into scheduling problems is given by Allahverdi et al. (1999) and Potts and Kovalyov (2000).

Some authors use a broader perspective for the lot-sizing phase and incorporate the machine assignment problem. As such, they define a production lot as a number of product units successively produced on a machine. A lot-sizing approach, so understood, is closely related to batching. However, in general, the point of view is different. A lot-sizing phase has demand volumes of different periods as input data, and an according procedure splits or combines these volumes to determine (production) lot-sizes. On the other hand, a batching routine has an intra-period, scheduling point of view. It combines (i.e. batches) jobs of the same product of a single period and assigns them to a machine. Consequently, the objectives of a lot-sizing and a batching procedure are also different. Lot-sizing procedures usually try to minimize costs, while batching routines follow non-cost and more scheduling-related objectives, such as flow time or makespan minimization.

Figure 1.1 shows the relationship between lot-sizing and scheduling as they are understood in this study. Three products a, b and c have to be produced on two machines in four periods. While lot-sizing is concerned with the product quantities, scheduling considers the specific machines and product sequences.

Lot-sizing and scheduling decisions are interdependent. A lot-sizing procedure calculates how many units of a product to produce in a period. To decide this, it needs information about the setup times. The setup times are determined by the machine assignment and product sequence, which are fixed by the scheduling procedure: If setup times are per se sequence-dependent, the relationship is obvious. However, the relationship holds true even if setup times are sequence-independent, as the last product of a period may often be continued in the next period without incurring an additional setup. Thus, the lot-sizing procedure needs an input from the scheduling procedure. On the other hand, the scheduling procedure needs the production volumes from

- Lot-Sizing: What quantities in which period?

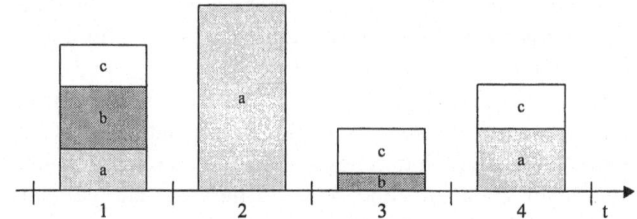

- Scheduling: Which machine and sequence?

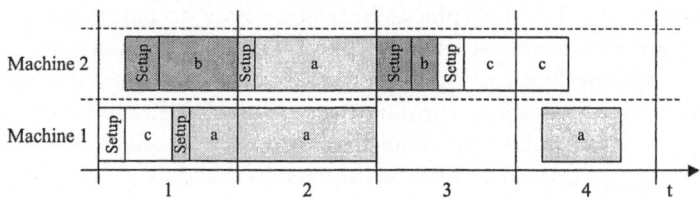

Fig. 1.1. Lot-sizing and scheduling

the lot-sizing phase in order to find an assignment and a sequence. As an example, consider the second period in Fig. 1.1. Machine 1 produces product a with a setup performed in period 1. The given production volume could not be produced if product a were not the last product on machine 1 in period 1. Thus, the product sequence is needed for the lot-sizing phase, but is available only after the scheduling phase. An integrative solution procedure that simultaneously solves the lot-sizing and the scheduling problem is needed to coordinate these interdependencies.

As pointed out by Drexl et al. (1994) and Tempelmeier (1997), the characteristics of lot-sizing and scheduling problems are different for each kind of production facility and specific solution procedures are needed. We will focus on a special kind of production facility: a flexible flow line.

1.2 Flexible Flow Lines

A flexible flow line—also commonly referred to as hybrid flow shop, flow shop with parallel machines or multiprocessor flow shop—is a flow line with several parallel machines on some or all production stages. Thus, all products follow the same linear path through the system. They enter it on the first stage and leave it after the last. On each of the stages, one of the parallel machines has to be selected for production. No two machines can simultaneously produce a product unit. The machines of a stage are identical, meaning that all machines of a stage can produce all products with the same process time. However, the

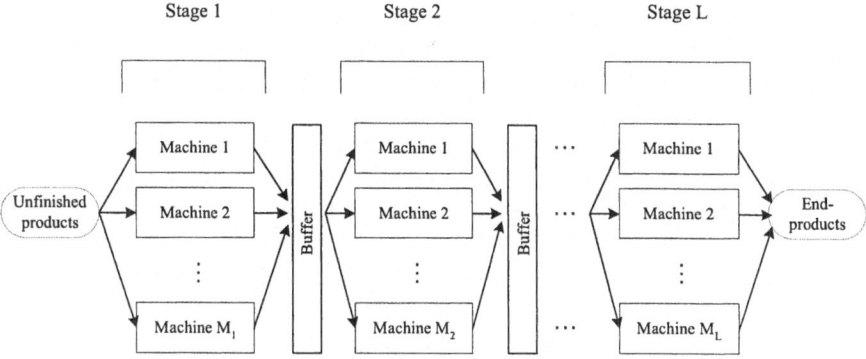

Fig. 1.2. Schematic view of a flexible flow line

process times may vary for different products and production stages. Hence, to reach similar production rates among stages, the number of machines per stage may be different as well.

Figure 1.2 shows a schematic view of a flexible flow line with L production stages and M_l machines on stage l. Between stages, buffers are located to store intermediate products.

While the term 'hybrid flow shop' is often used synonymously to 'flexible flow line', it is important to notice that the former is also used for other types of flow lines. For example, Tsubone et al. (2000) present a case study that includes parallel machines, but each machine can only produce one specific product. Obviously, such a production facility is quite different from a flexible flow line as discussed here.

Flexible flow lines can be found in a vast number of industries. They are especially common in the process industry. Numerous examples are given in the literature, including the automotive, chemical, cosmetics, electronics, food, packaging, paper, pharmaceutical, printing, textile and wood-processing industries (see Salvador 1973; Wittrock 1988; Adler et al. 1993; Guinet and Solomon 1996; Agnetis et al. 1997; Riane 1998; Botta-Genoulaz 2000; Moursli and Pochet 2000; Negemann 2001). We will investigate an application in the semiconductor industry.

1.3 Characteristics of the Problem

We consider a lot-sizing and scheduling problem for flexible flow lines, i.e. we have to determine production quantities, machine assignments and product sequences. A typical situation is as follows: A fixed planning horizon consists of one to several weeks and is divided into daily to weekly periods. Multiple products have to be produced, and there is a deterministic, discrete demand volume per product and period. The demand volumes vary for different periods because they mainly originate from customer orders. Moreover, some products may be at the end of their life-cycle and therefore show a decreasing demand while others are newly introduced to the market and thus show an increasing demand. All input materials required for production are available at the beginning of the planning horizon.

We have identified several important features in industrial practice that are relevant for lot-sizing and scheduling of flexible flow lines. Mainly, these are (1) the inclusion of back-orders, (2) the aggregation of products to product families and (3) the explicit consideration of parallel machines.

It is possible to produce a product volume before its demand period. In this case, it is stored in inventory, where it remains until its delivery to the customer in the demand period. For every period and unit of inventory, product-specific inventory costs are incurred. It is also possible to produce a product after its demand period. In this case, the product is tardy, and associated **back-order** costs are incurred for every unit and period of the delay. The precise production time within a period is irrelevant for inventory or back-order costs. While most research neglects the possibility of back-ordering, it is of great importance in practical settings: If capacity is scarce, some products may *have* to be back-ordered. Hence, it is not relevant to come to the result that the demand volume cannot be produced on time. In fact, the question is, which demand volumes shall be back-ordered and which shall not? This question is solved by explicitly incorporating the possibility of back-orders into the problem formulation.

Since the production facility is arranged as a (flexible) flow line and all products follow the same linear path through the system, the products are relatively similar. There is, however, an increasing number of variants that has to be produced simultaneously (see Thonemann and Bradley 2002). Thus, it is typical that certain subsets of variants (products) can be grouped to **product families**. The variants differ only slightly. It is therefore reasonable to assume that the products of a family share the same inventory, back-order and setup costs. Further, for any given machine, they have the same process time. Hence, from a logistical point of view, the products of a family are identical. However, when a machine changes production from one product variant to another, a setup has to be performed.

We consider two kinds of setups. *Setups between product families* incur a setup time and setup costs. Setup costs may consist of material or labor costs, but not of the costs associated to lost machine capacity. Lost machine capacity is explicitly incorporated through setup times. We assume that product family setups are sequence-independent and identical for all parallel machines of a production stage. Thus, setup costs and times only depend on the production stage and the product family that a machine is being set up to. Even if setups are sequence-dependent, this approach is still applicable if the degree of sequence-dependency is moderate and/or outweighed by other (stochastic) factors. In this case, it is often possible to use average setup times and costs. *Setups within a product family* (intra-family setups) incur setup costs, but it is assumed that the respective setup times are negligible. Intra-family setup

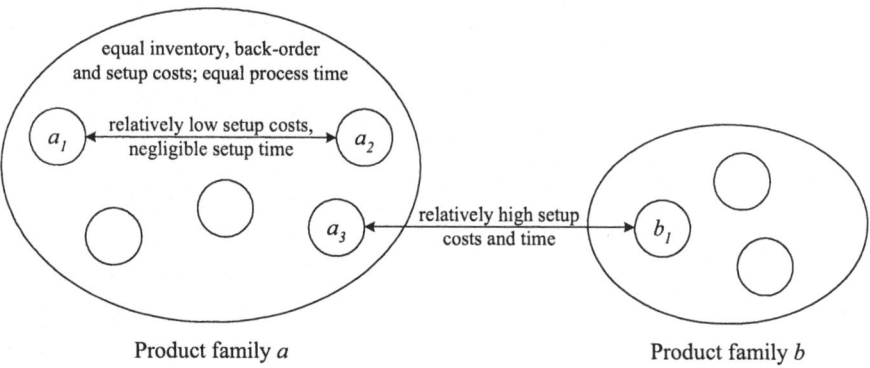

Fig. 1.3. Relationship between products and product families

costs are assumed to be identical for all products of a family and all parallel machines of a production stage. Since the products of a family are very similar, setup costs between families are higher than intra-family setup costs. Hence, there is some kind of sequence-dependency: When setting up a machine to a product of family a, it makes a difference if the machine has been set up for another product of family a or for a product of family b. In the latter case, setup costs will be higher and also setup times may be incurred. Figure 1.3 summarizes the relationship between products and product families.

The production facility is arranged as a flexible flow line. This implies that there are **parallel machines** on some or all production stages. The demand volume of a product can be divided into several product units. The product units may be produced on any of the parallel machines. Hence, we have to determine on which machine to produce each product unit and how many machines to set up for a product. In the remainder, a product unit will also be called a 'job'. A job consists of several 'operations', one at each production stage. When an operation is started on a machine, it will be finished without interruption. Thus, pre-emption is not allowed. Each product has to visit all production stages.

Generally, the buffer space between stages (as well as before the first and after the last stage) is limited. However, infinite buffers may mostly be assumed because the product units are sufficiently small. We further assume negligible transport times between stages, which is reasonable because of the flow line characteristics of the production facility: In practical settings, process times are usually much longer than transport times. We require that machines are available all the time, i.e. they never break down. However, machine breakdowns may be incorporated on an average basis by a prolongation of process times. Defective parts and rework are not considered. All data is assumed to be deterministic.

Figure 1.4 shows an Entity-Relationship Model of the problem (for an introduction to data modeling using Entity-Relationship models, see e.g. Batini

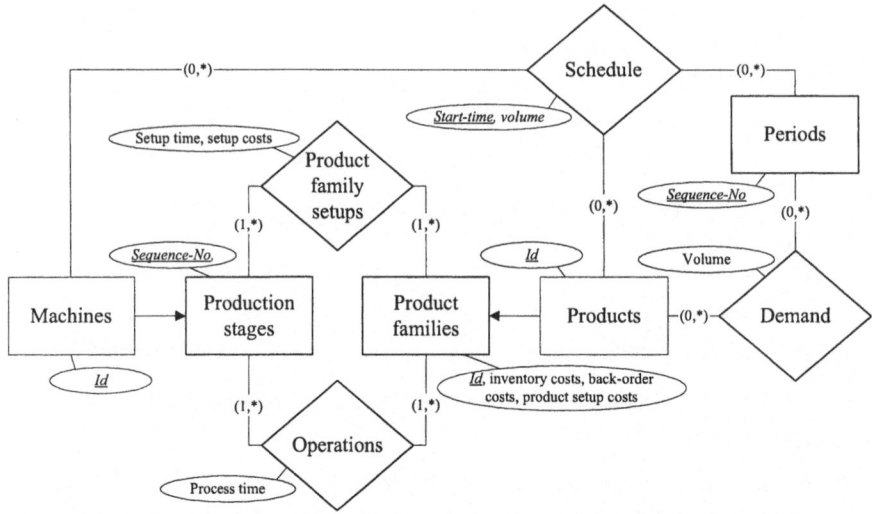

Fig. 1.4. Entity-Relationship Model of the problem

et al. 1992). The 'schedule'-entity represents the solution of the problem. It combines the information on the lot-sizes (given by its 'volume'-attribute) with the ones on the schedule (given by the relationship to the 'machines'-entity and the 'start-time'-attribute) and thus delivers a solution to both problems as depicted in Fig. 1.1.

A typical objective of lot-sizing procedures is to minimize all related costs. In our case, these are inventory holding, back-order and setup costs. Scheduling procedures usually have a time-oriented objective. When there are no explicit due dates within the planning period, one approach is to minimize the completion time of the last operation (makespan). This implies minimizing the total time needed for production, and thus to produce the given production volume as quickly as possible. When the jobs have to be finished at a certain time during the planning period, a direct measure of due date fulfillment is often used as the objective. This implies minimizing the (weighted) tardiness of jobs. Some approaches also try to avoid job earliness. When jobs consist of multiple operations—such as in a flexible flow line—a common approach is to consider the flow time. A job's flow time is the time between the beginning of its first operation until the completion of its last operation. In the case of a flexible flow line, this is a job's production start on the first stage until its completion on the last stage. The objective of a low flow time implies that the waiting time between operations shall be minimized. Thus, it minimizes work in process (WIP) and associated inventory costs as well as buffer usage.

In our case, due dates are considered by inventory holding and back-order costs between periods. There are no due dates within a period. Hence, the intra-period objective is to produce a period's production volume as efficiently as possible. Since we consider a multi-stage system, the minimization of a job's flow time seems an appropriate measure. Therefore, the objective used in this study is to find a schedule that

1. minimizes total costs (primary objective, for the lot-sizing problem) and
2. minimizes the mean flow time through the system (secondary objective, for the scheduling problem).

The two objectives contribute favorably to a high capacity utilization (because the minimization of job waiting times implies low machine idle times) as well as fulfillment of the delivery dates (products are produced in a period that minimizes the related inventory holding and back-order costs, and once production is initiated, a job's waiting time is minimized). Our notion of mean flow time is equivalent to the 'average interval' or 'sojourn' time as introduced by Kochhar and Morris (1987). They argue in favor of this objective because it also leads to higher and more controllable quality—for example, because manufactured products quickly reach an inspection stage and process defects may be corrected rapidly.

1.4 Scope and Outline of the Study

In recent years, hierarchical planning approaches have been adopted by several commercial vendors of production planning systems. Examples are SAP with 'Advanced Planner&Optimizer', Manugistics with 'Supply Chain Management' or i2 Technologies with 'i2 Six'. Due to the increasing necessity of such systems, the revenues of these companies have risen sharply over the last few years (Meyr 1999, Fleischmann and Meyr 2003).

According to the hierarchical approach, these planning systems divide the overall problem into several phases, and the resulting sub-problems are solved using specialized procedures. However, the algorithms embedded in those systems often lack quality and applicability to real business problems (Tempelmeier 2003, p. 350). For example, they may neglect important constraints (like capacity restrictions) or do not consider special relationships (like multiple production stages).

The study at hand offers a new integrative solution procedure for the lot-sizing and scheduling problem for flexible flow lines. Thus, it may help improve production planning systems by delivering a specialized routine for one of the planning phases.

The approach may be employed to solve real business problems in basically all industries that use flexible flow lines with the above characteristics. It has been successfully employed in the semiconductor industry.

The outline of the study is as follows: Chapter 2 contains a literature review on lot-sizing and scheduling procedures for flexible flow lines. There seems to be no literature on flexible flow lines considering lot-sizing and scheduling problems simultaneously. Chapter 3 presents an integrated solution approach that is capable of incorporating the interdependencies of the two problems. It consists of three phases, which are described and tested in Chaps. 4, 5 and 6. An illustrative example covering the complete algorithm is presented in Chap. 7. Chapter 8 portrays an application of the approach in the semi-conductor industry, where it has also been employed for a capacity planning study. We summarize the results and draw conclusions in Chap. 9.

2

Lack of Solution Procedures in the Literature

Time is a one-way street!

To our knowledge, there is no literature combining lot-sizing and scheduling decisions for flexible flow lines in the sense defined above. A paper by Lee et al. (1997)—called 'A genetic algorithm-based approach to flexible flow-line scheduling with variable lot sizes'—neither covers parallel machines (the authors do not define their understanding of a 'flexible flow line'), nor does it solve lot-sizing decisions in the sense of this study. Using our notation, they solve a scheduling and batching problem for standard flow lines with a single machine on each stage.

So far, the lot-sizing and the scheduling problem for flexible flow lines have only been considered separately. Such dichotomic approaches cannot coordinate the interdependencies of the two problems. Research on stand-alone lot-sizing problems is also very limited. We will present two papers in Sect. 2.1. Related papers covering lot-sizing problems in adjacent areas are reviewed in Chap. 4. There is, however, a rich body of literature considering stand-alone flexible flow line scheduling problems. Research on this area is classified and discussed in Sect. 2.2. We concentrate on the case where preemption is not allowed.

Only a few studies consider setup times between jobs, especially studies concentrating on the scheduling problem neglect setups. They usually assume that all jobs belong to different products. In this case, a job's setup time can be added to its process time (provided setup times are sequence-independent). However, in our case, setup times (and costs) are important. Hence, we explicitly mention their inclusion in a study.

Generally, we have to distinguish between optimal solution procedures and heuristics. Optimal solution procedures find the (or a) best solution for a given criterion. The drawback of optimal solution procedures is their computation time. For a class of difficult—so-called NP-hard—problems, one cannot expect to develop an optimal solution procedure that solves medium-sized or large instances in a reasonable amount of computation time (see Garey and Johnson 1979). In contrast, heuristics do not (provably) deliver optimal, but mostly adequate solutions in a moderate amount of time. Some of them are applicable for medium-sized and even large problems as they appear in real business applications.

2.1 Lot-Sizing Problems

To our knowledge, there are only two studies considering (stand-alone) lot-sizing problems for flexible flow lines.

Derstroff (1995) considers a multi-level job shop problem and extends it to include alternative routings on parallel machines. While in job shops, arbitrary routes through the system are possible, a flexible flow line may be interpreted as a special case of such a system—if the machines can be clustered to form a production stage and all products follow the same linear path through the system. Derstroff solves the problem using a heuristic Lagrangian relaxation procedure. Özdamar and Barbarosoglu (1999) consider a flexible flow line lot-sizing problem. However, so-called 'lot spitting' is not allowed: If a product is produced in a period, the complete production volume must be assigned to a single machine. The problem is solved heuristically. Both studies consider setup times. Further details are presented in Chap. 4.

The presented approaches consider the stand-alone lot-sizing problem. Thus, they determine the production volume per product and period. They also assign a machine to each product unit. However, they do not sequence the products on a machine and therefore neglect the interdependencies between lot-sizing and scheduling.

2.2 Scheduling Problems

Scheduling procedures for flexible flow lines can be categorized according to their solution approach.

One stream of research is focussed on optimal solution procedures. Scheduling flexible flow lines is an NP-hard problem for all traditional optimality criteria such as minimizing the makespan, even when setup times are negligible (see Garey et al. 1976; Kis and Pesch 2002). Thus, the computation time of optimal solution procedures is prohibitively long when medium-sized or large instances are considered. However, a number of optimal solution procedures have been developed for the flexible flow line scheduling problem. They can be used for rather small problem instances. Most of them are based on Branch&Bound algorithms.

Another stream of research tackles flexible flow line scheduling problems in a heuristic way. Local search and metaheuristics are one group of such procedures. Another group decomposes the problem. Three sub-classes can be identified: Stage decomposition approaches, job-oriented procedures, and algorithms that solve the loading and sequencing part sequentially. One paper in the last category iteratively solves the two components and finds an optimal solution to the original problem.

Each of the following sections discusses one of these approaches and presents related literature.

2.2.1 Branch&Bound Procedures

Branch&Bound is a standard technique for combinatorial optimization problems. A general description is for example given by Michalewicz and Fogel (2000) or Blazewicz et al. (2001). The basic idea is as follows: The original problem is simplified by relaxing some constraints that make it difficult. Sub-problems are generated by iteratively adding constraints and fixing decision variables. The sub-problems are organized in a tree-like structure. While in the worst case, all possible combinatorial solutions must be evaluated, the tree-structure usually allows the neglect of some solutions without explicitly considering them. This is made possible by lower and upper bounds to the objective value of the original problem: When it is known that all solutions of a branch cannot be better than the current best solution, the branch can be deleted ('fathomed').

The bounds are improved during the search process, and the generation of good bounds is crucial for Branch&Bound procedures. When the bounds are not tight, many solutions must be evaluated, leading to long computation times. In any case, as mentioned above, Branch&Bound procedures need a relatively long computation time, which limits their applicability to small problems. However, they may be terminated prematurely as soon as a first solution has been found, which would make them a heuristic.

Branch&Bound procedures for a specific problem differ, mainly, in four ways: The constraints that are being relaxed, the way that sub-problems are created ('branching scheme'), the order in which the sub-problems are considered ('search strategy'), and the creation of bounds. A review of Branch&Bound procedures for scheduling flexible flow lines is given by Kis and Pesch (2002).

None of the Branch&Bound procedures presented in this section includes setup times. Furthermore, most procedures focus on minimizing the makespan. Thus, unless otherwise noted, makespan minimization is the objective of the studies presented in this section.

Salvador (1973) was among the first to consider flexible flow line scheduling problems in the literature. His Branch&Bound procedure generates a permutation schedule, i.e. the same sequence of jobs is used on all stages. A permutation schedule simplifies the problem substantially, as only one sequence of jobs has to be determined. However, non-permutation schedules may result in a better makespan. In the considered problem, in-process inventory is not allowed.

Brah and Hunsucker (1991) develop a Branch&Bound algorithm that is able to generate non-permutation schedules with machine idle times between jobs. For a particular machine, a schedule is given by a sequence of jobs. A schedule for the flexible flow line can be derived by a concatenation of such sequences, one for each parallel machine on all production stages. The search tree consists of such concatenated sequences. Thus, the problem is treated on a stage-by-stage basis. However, the computation time of the suggested procedure is rather long.

Rajendran and Chaudhuri (1992b) consider permutation schedules and solve the problem by generating a single sequence of jobs that is valid for all production stages. They exploit the fact that the makespan will be identical for a number of special sub-sequences, which allows them to neglect parts of the solution space.

Brockmann et al. (1997) develop a parallelized version of the algorithm presented by Brah and Hunsucker (1991), which can be used on parallel computer processors to speed up computation times. Brockmann and Dangelmaier (1998) improve the algorithm by using a different branching scheme and adjusted lower bounds. In the latter version, the algorithm also covers machine availability times. Portmann et al. (1998) enhance the approach of Brah and Hunsucker (1991) by coupling it with a Genetic Algorithm, which is employed to derive upper bounds (i.e. schedules) during the Branch&Bound procedure. They also use tighter lower bounds.

Moursli and Pochet (2000) use a branching scheme that is an extension of a method for a parallel machine scheduling problem. The schedule is generated one stage at a time. An approach based on Branch&Bound coupled with constraint propagation is presented by Carlier and Néron (2000). Unlike the above mentioned procedures, it considers the stages simultaneously. The search tree is generated by consecutively selecting a stage and the next job to be scheduled on that stage. Néron et al. (2001) use the so-called concepts of 'energetic reasoning' and 'global operations' to reduce the computation time of the procedure.

Research on Branch&Bound procedures with objectives other than make-span minimization is sparse. Rajendran and Chaudhuri (1992a) consider the objective of minimizing mean flow time. In their research, flow time refers to the time from the beginning of the planning interval until a job's completion time on the last stage (in contrast to the time between a job's production start on the first stage to its completion on the last stage, as it is used in the present study). They develop a Branch&Bound procedure similar to the one by Rajendran and Chaudhuri (1992b). It generates an optimal permutation schedule. Azizoglu et al. (2001) consider the same problem, but allow non-permutation schedules. They compare the branching scheme of Brah and Hunsucker (1991) with one that does not generate certain sub-problems because of proven sub-optimality under the mean flow time objective. With the new branching scheme, the algorithm has a substantially lower computation time.

2.2.2 Local Search and Metaheuristics

Local search procedures and metaheuristics are based on an initial solution—which may be generated by an easy and straightforward approach. They try to improve the initial solution by consecutively changing small parts of it. For each current solution, a number of 'neighboring' (i.e. similar) solutions are generated and evaluated—which is why these heuristics are also called 'neighborhood search procedures'. The term 'local search' is sometimes used to describe a somewhat myopic variant that only accepts a new solution if it is better than the preceding one. If no better solution can be found, the procedure terminates. To overcome local optima, 'metaheuristics' may be employed. Metaheuristics also accept solutions that are worse than the current one, in the hope of reaching other regions of the solution space with better solutions than the best one so far. An example representative of metaheuristics is 'Tabu Search', which—in a simple version—accepts all solutions that have not been visited recently (and thus marked tabu). Most metaheuristics incorporate a random element in the search process. For example, 'Simulated Annealing' only accepts a worse solution at a given probability that depends on the degree of the deterioration and the number of iterations so far. The probability of accepting a worse solution decreases ('anneals') with the number of iterations. In Genetic Algorithms, the creation of new solutions is random by itself. They operate on a pool of current solutions. Two solutions ('father' and 'mother') are selected at random and a new ('offspring') solution is generated by copying parts of the parent solutions. Genetic Algorithms will be discussed in Chap. 6. Further details about local search procedures and metaheuristics in general are given by Reeves (1995), Osman and Laporte (1996), and Michalewicz and Fogel (2000).

In flexible flow lines, a job consists of several operations, one for each production stage. Thus, a flexible flow line schedule consists of machine assignments for each operation as well as a production sequence for each machine. A move to a neighboring schedule may change the machine assignment of an operation or its position in the sequence. This neighborhood is very large. Hence, local search procedures and metaheuristics must find ways to limit the size of the neighborhood. This can be done by only allowing certain moves that appear promising.

Negemann (2001) compares several Tabu Search heuristics with a Simulated Annealing procedure. Among the Tabu Search heuristics are approaches by Hurink et al. (1994), Dauzère-Pérès and Paulli (1997), as well as Nowicki and Smutnicki (1998). The objective is to minimize makespan. He concludes that the neighborhood of Nowicki and Smutnicki performs best. This neighborhood is tailor-made for flexible flow lines, and only allows moves that have a direct impact on the makespan.

A completely different approach is adopted by Leon and Ramamoorthy (1997). Instead of considering neighborhoods of a schedule, they consider neighborhoods of the problem data. The main idea is that a slight variation of the input data will still result in good solutions for the original problem. Since every heuristic is somewhat myopic, it might even be that the perturbed data will lead to better solutions for the original problem than the original problem data itself. To generate a schedule, the authors employ a simple procedure that schedules one operation at a time. The procedure is invoked for each input data set. A local search procedure is used to coordinate the creation of input data sets. The objective is to minimize makespan or mean tardiness of jobs. Jobs are allowed to skip stages. Kurz and Askin (2001) report on the good solution quality of the algorithm.

Some of the algorithms discussed in the following sections also incorporate local search procedures and metaheuristics. However, in these sections, the latter are superposed by other fundamental concepts of the respective algorithm.

2.2.3 Stage Decomposition Approaches

The main idea of stage decomposition approaches is to split a flexible flow line along the production stages. This leads to several single stage, parallel machine scheduling problems, each of them with reduced complexity compared with the overall problem. Single stage parallel machine scheduling problems have received a lot of attention in the literature. Mokotoff (2001) gives an overview.

We call a procedure a stage decomposition approach if the following two criteria are satisfied:

1. The stages are scheduled separately by priority rules (so-called 'list schedules' as introduced by Graham (1966 and 1969)).
2. The generation of a schedule is based on a sequence of jobs, which is used
   to schedule the first production stage.

While the first criterion is necessary for stage decomposition approaches,
a procedure may also be categorized as a stage decomposition approach if the
second criterion is not satisfied. For example, a simple stage decomposition
approach is to use dispatching rules on every stage to select the next job that
has to be produced whenever a machine becomes idle. Agnetis et al. (1997)
use such a procedure ('pull rule') in the automotive industry, where special
constraints—such as the availability of automated guided vehicles—prevent
more sophisticated algorithms from being employed without major modifications. They evaluate various objective criteria. Adler et al. (1993) focus on
the bottleneck instead of the first production stage as implied by the second
criterion above. They consider an application in the packaging industry with
sequence-dependent setup times. The objective is to minimize mean tardiness
and work in process as well as to maximize throughput. The solution approach
schedules the bottleneck stage by a priority rule. The resulting sequence of
jobs is used to schedule the other stages. Some local re-sequencing is allowed
to save setup times.

A problem with identical jobs and makespan minimization is presented by
Verma and Dessouky (1999). The production stages are scheduled separately,
and miscellaneous priority rules are compared. Because all jobs are identical,
the sequence of jobs is irrelevant. For general problems with unidentical jobs,
the sequence of jobs has to be considered, because a sequence that is good
for one production stage may be bad for another. Hence, not considering the
stages simultaneously will lead to sub-optimal solutions. Stage decomposition
approaches may be classified by the way they try to overcome this problem:
One approach is to employ techniques based on standard flow line algorithms.
Another one is to use local search or metaheuristics, i.e. to generate and
evaluate more than one initial job sequence. The remainder of this section
discusses both approaches.

2.2.3.1 Standard Flow Line Adaptations

We define a 'standard flow line' as a flow line with a single machine on each
stage. There is plenty of research on scheduling such systems. Cheng et al.
(2000) provide an overview. The heuristics discussed in this section use a
standard flow line algorithm to incorporate the dynamics and relationships of
multi-stage environments into a stage decomposition approach.

All standard flow line adaptations follow a certain framework: In a first
step, the heuristics aggregate the machines of a stage to a single, fictitious machine. This, for example, can be done by dividing a job's process time by the
number of parallel machines. Afterwards, a job sequence for the first stage

is generated. This is the task of a technique originally developed for standard flow lines—using the aggregated, fictitious machines. In a second step, the parallel machines are considered explicitly. A loading procedure selects a machine for each job, in the order of the job sequence from the first step. Subsequent production stages are scheduled consecutively by taking the resulting sequence of a previous stage as input. This is done using priority rules that may be derived from single stage parallel machine scheduling algorithms. A different rule may be employed for each production stage.

Thus, the flow line characteristics of the problem are taken into account by the sequence in which the jobs are being considered on the first production stage. The heuristics are based on the premise that such a sequence will lead to good overall schedules on all stages. The scheme is quite general and can be used for different objectives. Several heuristics have been developed according to the above framework. Their main differences are based on:

1. a different method to aggregate the machines of a production stage in the first step.
2. a different procedure to generate the job sequence for the first stage. Basically every heuristic for standard flow lines may be used. For makespan minimization, often a modification of Johnson's rule (Johnson 1954) or its extension by Campbell et al. (1970) is employed.
3. a different assignment procedure in the second step.
4. different priority rules to schedule later production stages.

Ding and Kittichartphayak (1994) develop three variants of flow line adaptations for makespan minimization. Guinet and Solomon (1996) consider a problem with the objective of minimizing the maximum tardiness and compare several variants. A comparative study for makespan and mean flow time minimization is conducted by Brah and Loo (1999). A study by Botta-Genoulaz (2000) incorporates setup times and job precedence constraints. Several heuristics are investigated under the objective of minimizing the maximum tardiness.

Kurz and Askin (2003a) compare several methods for a makespan minimization problem with sequence-dependent setup times. Jobs are allowed to skip stages. Three groups of heuristics are evaluated: A first group is based on simple assignment rules. A second group is based on insertion heuristics for the Traveling Salesman Problem in order to take the sequence-dependent setup times into account. A third group is based on Johnson's rule. The latter outperforms the others.

Koulamas and Kyparisis (2000) develop a linear time algorithm for two- and three-stage flexible flow lines. The procedure is (with an increasing number of jobs) asymptotically optimal for makespan minimization. A three-stage flexible flow line under the same criterion is also considered by Soewandi and Elmaghraby (2001). They develop four variants of standard flow line adaptations.

2.2.3.2 Entry Point Sequencing via Local Search

Some authors have used a different approach to incorporate flow line characteristics into stage decomposition approaches. Instead of (solely) depending
on a standard flow line technique to sequence the jobs for the first production
stage, they employ local search procedures or metaheuristics that create and
successively improve the sequence. The sequence for the first production stage
is called 'entry point sequence'. The schedule itself is generated analogously to
the second step of standard flow line adaptations. Thus, a loading procedure
selects a machine for each job as given by the entry point sequence. Subsequent stages are scheduled consecutively, using priority rules. The complete
schedule is generated for each candidate sequence and an evaluation criterion
is computed. Based on the criterion, the local search procedure or metaheuristic decides whether or not to accept the new solution and proceeds with its
operation. The difference to the local search procedures and metaheuristics
presented in Sect. 2.2.2 is that the procedures here are based on a job sequence
for the first production stage. Neighboring solutions are permutations of the
current entry point sequence. The algorithms presented in Sect. 2.2.2 use a
different representation format.

The term 'entry point sequence' was introduced by Kochhar and Morris (1987). Their study incorporates setup times and limited buffers between
stages. The objective is to minimize the average time a job spends between
entering the first and leaving the last stage. Starting with a random sequence,
two different neighborhood-operators are applied iteratively until no further
improvement can be achieved. One operator has a tendency to create batches
of jobs belonging to the same setup state, while the other one splits up the
batches. Kochhar and Morris discuss several priority rules to load the parallel
machines of each stage.

Riane (1998) determines the entry point sequence using a Simulated Annealing approach. The first stage is loaded according to the entry point sequence, later stages are loaded according to the release times imposed by
the respective previous stage. The loading algorithm is able to handle setup
times. The objective is to minimize makespan. Riane et al. (2001) embed the
approach in a decision support system for production planning and scheduling.
Lot-sizing decisions are not part of the described planning process.

Jin et al. (2002) consider a flexible printed circuit board assembly line,
also under makespan minimization. In their approach, a Genetic Algorithm
is used to generate a good job sequence. The starting point is a sequence
generated by a standard flow line adaptation. The procedure is presented for
three production stages, but could easily be extended to incorporate further
stages. Another Genetic Algorithm is developed by Kurz and Askin (2003b).
Their approach includes sequence-dependent setup times. Jobs do not have to
visit all production stages, and the objective is makespan minimization. When
more than two stages are considered, the Genetic Algorithm outperforms the
procedures presented by Kurz and Askin (2003a).

Wardono and Fathi (2003a) employ a Tabu Search algorithm. Again, the objective is makespan minimization. The algorithm leads to similar results to the one by Nowicki and Smutnicki (1998) presented in Sect. 2.2.2. Wardono and Fathi (2003b) extend the heuristic to cover limited buffers between production stages. Numerical results indicate that it outperforms procedures by Sawik (1993) and Wittrock (1988), which will be presented below.

2.2.4 Job-Oriented Procedures

Job-oriented procedures consider the jobs subsequently. In each iteration, a job is selected and loaded on a machine on each production stage. Sequencing decisions are made simultaneously.

Sawik (1993) develops a job-oriented heuristic for a flexible flow line with limited buffers between stages. The objective is to minimize makespan. In each iteration, a job is selected and loaded (appended) on the earliest available machine on each stage. The goal of the job selection is to minimize machine waiting and blocking times, with a priority for jobs with a long process time on downstream stages. Later, Sawik (1995) extends the heuristic for the case of no buffers between stages.

Gupta et al. (2002) consider a problem where some of the process times are not fixed, but have to be determined between a minimum and a maximum value. The optimization criterion is based on due dates. An initial job sequence is generated using priority rules. Considering the jobs in this order, they are inserted at the best possible position of a second job sequence. Thus, instead of simply appending one job after the other, the best insertion position is selected. This is done using a scheduling method that generates a plan for all production stages by loading the jobs on the first available machine. A local search procedure concludes the algorithm. Based on the second job sequence, it iteratively shifts job positions and re-evaluates the schedule. The authors report that enhancing the local search by Simulated Annealing does not improve results.

Besides the 'pull rule' mentioned above, Agnetis et al. (1997) also consider a 'push rule': Whenever a job is finished on a production stage, the rule selects a machine for production on the subsequent stage. In this way, the method 'pushes' the jobs through the system, which can be considered a job-oriented procedure. With a push rule, a job may remain in a local buffer of a machine while another machine is idle. In contrast, a pull rule establishes a global buffer that feeds all machines of a stage. Therefore, the pull rule leads to better results. However, it may imply a modification of the facility layout.

2.2.5 (Sequential) Loading and Sequencing Algorithms

As pointed out by Sawik (2000), flexible flow line problems may be divided into a loading and a sequencing part.

Loading refers to the allocation of a job to a machine: For each job, one of the parallel machines has to be selected per production stage. When all loading decisions have been made, the routes through the system are determined. Thus, the problem reduces to a job shop scheduling problem, which constitutes the sequencing phase. In the sequencing phase, a production sequence for all jobs assigned to each machine has to be generated.

Solving these two problems subsequently reduces the overall complexity. However, their interactions cannot be taken into account and the results are generally inferior. Sawik (2002) gives an example using a comparison for a printed wiring board manufacturing facility.

Wittrock (1985) describes a similar practical case. The objective is to find a schedule that minimizes makespan and queueing in intermediate buffers. He presents a three-phased algorithm consisting of a machine loading procedure (assigning jobs to machines), a subsequent sequencing method (calculating the order in which jobs enter the system) and a final timing step (determining when the jobs enter the system). The result is a cyclic schedule, which is repeated periodically to produce the complete demand. Later, Wittrock (1988) drops the periodic arrangement and re-examines the sequencing method. Instead of a simple rule that consecutively appends jobs, he uses a heuristic based on Dynamic Programming techniques. Moreover, the algorithm is extended to handle limited buffers, urgent jobs, and initially non-empty lines. The ability to handle initially non-empty lines allows re-scheduling in response to unpredicted events such as machine breakdowns.

Brandimarte (1993) uses two nested Tabu Search procedures to solve the scheduling problem. The outer procedure considers the assignment problem, the inner one deals with the sequencing issues. The objective is to minimize makespan or total tardiness of jobs. The algorithm is included in the above mentioned comparative study by Negemann (2001).

Harjunkoski and Grossmann (2002) develop a method that generates an optimal solution by iteratively solving the loading and the sequencing problem consecutively. The objective is to minimize job assignment costs and one-time machine-initialization costs. Setup times are included, but are only dependent on the machine and not on the product. The algorithm is based on a hybrid mixed integer and constraint programming approach: The loading problem is solved using mixed integer programming. A subsequent constraint programming procedure solves the sequencing part and iteratively adds cuts to the loading problem until an optimal schedule is found.

2.3 Summary and Conclusion

The literature review has shown that on the one hand, there are stand-alone
lot-sizing procedures that determine production volumes per product and period and assign a machine to each product unit. They do not sequence the
products loaded on a machine. On the other hand, there are scheduling procedures that solve the machine assignment and the sequencing problem, but
do not merge demands of different periods to calculate lot-sizes. There seems
to be no study combining lot-sizing and scheduling decisions for flexible flow
lines. Such an integrative approach is needed to incorporate the interdependencies between the two problems. We develop such an approach in the next
chapter.

3

An Integrative Solution Approach

It's time to tune into... the Twilight Zone!

In the following, we illustrate an integrative lot-sizing and scheduling approach for flexible flow line production facilities.

The special characteristics of the problem have been addressed in Chap. 1. The objective is to find a feasible schedule that minimizes all relevant costs, which are inventory, back-order and setup costs. A secondary objective is to minimize mean flow time—i.e. the average time between a job's production start on the first stage until its completion time on the last stage. The overall problem is too difficult and complex to be considered in a monolithic way. Hence, we present a hierarchical approach that consists of three phases:

1. Bottleneck planning
2. Schedule roll-out
3. Product-to-slot assignment

The approach is bottleneck-oriented. A human planner is usually aware which production stage constitutes the bottleneck (see e.g. Adler et al. 1993 or Pinedo 2002). If this is not the case, the bottleneck stage can be determined by the incurred workload of the given demand volumes. Generally, the stage with the highest workload is the bottleneck stage. However, two additional factors have to be considered: Firstly, setup times have an influence on the workload, but they are a result of the planning procedure. Thus, to determine the bottleneck, the sum of setup times has to be estimated. If several stages are candidates for the bottleneck, the one with the longest average setup time per product family should be chosen. This stage is the most inflexible and should be focused on in order to find a good, feasible solution. Secondly, if two stages are similar even after considering setup times, the most downstream stage should be selected. This is because this stage has the most direct impact on the completion time of jobs, and hence on the delivery times to the customer. In the remainder, we assume that the bottleneck stage is known beforehand.

Figure 3.1 depicts the phases of the solution approach along with their functions in the overall context, the level of detail, the considered production stages, and the objectives. With the initial focus on the bottleneck stage,

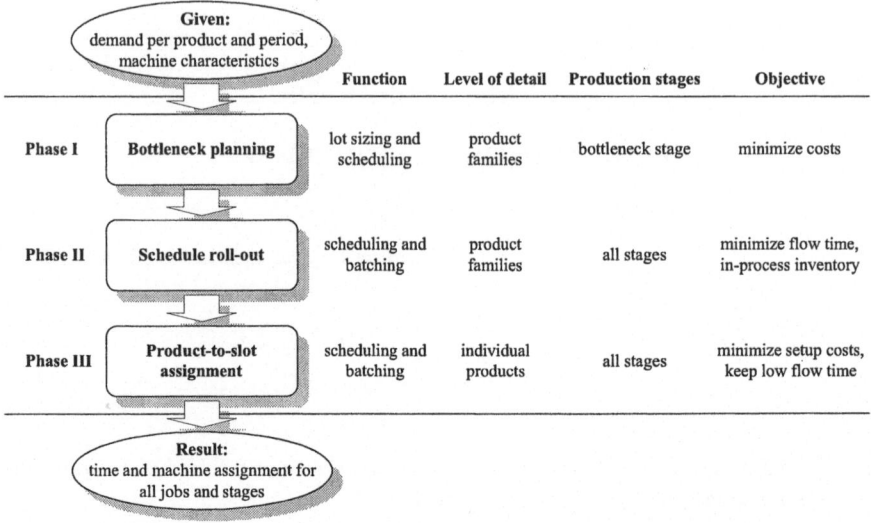

Fig. 3.1. Lotsizing and scheduling approach for flexible flow lines

the solution approach decomposes the flexible flow line along the production stages. The bottleneck stage is planned separately from the non-bottleneck stages, which are, afterwards, all considered simultaneously. However, when scheduling the bottleneck stage, the other stages are taken into account implicitly in order to find a good overall schedule. The first two phases work on an aggregation level of product families. A schedule for individual products is generated in the third phase. Hence, the approach is conceptually based on a decomposition and aggregation strategy. Figure 3.2 shows the decomposition and disaggregation steps with an example covering three product families A, B and C. The initial situation with the given demand volumes for three periods is illustrated in the top panel.

Phase I considers the decomposed and aggregated problem, which is a single stage lot-sizing and scheduling problem on product family level for the bottleneck stage. In order to incorporate the dependencies between the lot-sizing and scheduling decisions, the two have to be considered simultaneously. We determine production volumes, assign machines, and sequence the product families on the bottleneck stage. An example is shown in the second panel of Fig. 3.2.

The resulting production quantities of a period will not only be used on the bottleneck stage, but also on the other production stages. Thus, the other phases do not make lot-sizing decisions and Phase I pre-determines the completion time of end-products. As a consequence, the bottleneck planning phase is responsible for the inventory and back-order costs of end-products. It is also possible to integrate system-wide setup costs into the bottleneck planning phase: The objective of the second phase is to minimize flow time. To

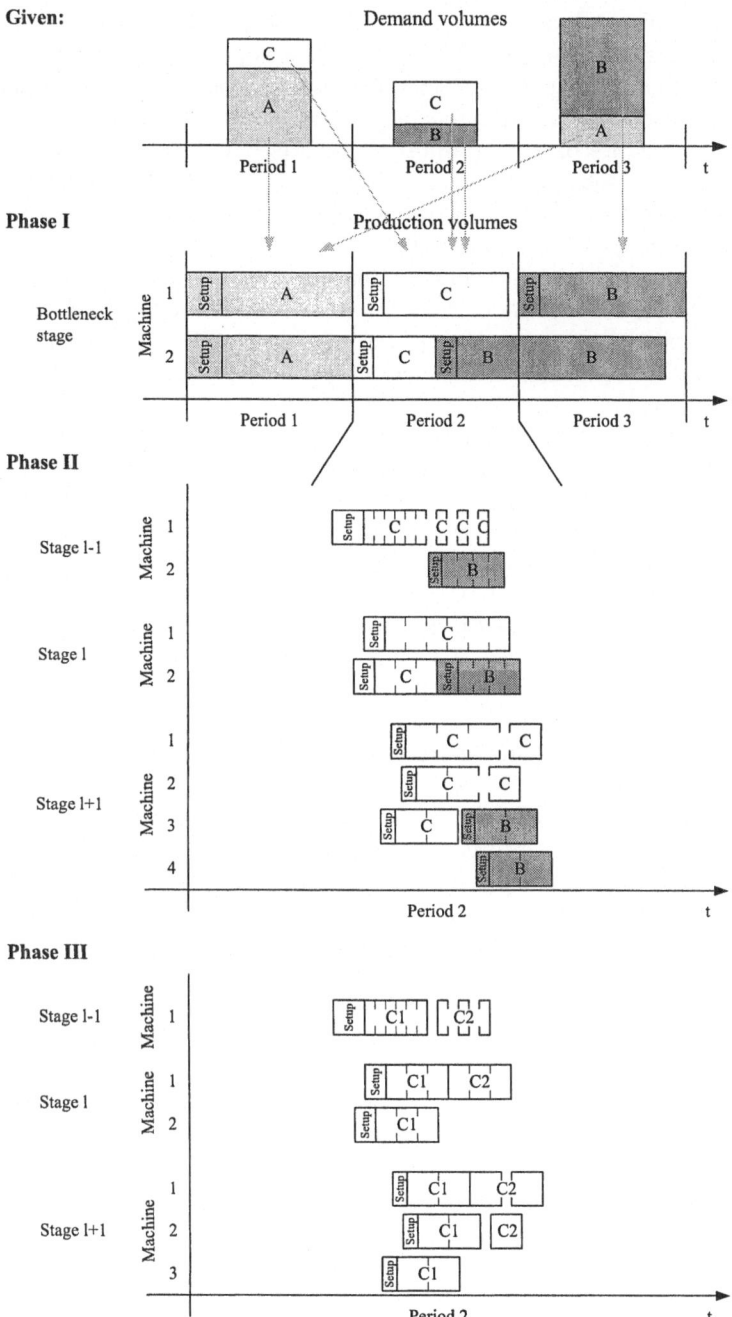

Fig. 3.2. Decomposition of the problem into three planning phases at different aggregation levels

accomplish this goal, a job has to be started as soon as it is completed on a previous stage. The number of parallel machines producing a product family on the bottleneck stage determines how many jobs are produced per time unit. The more machines produce a product family on the bottleneck stage, the more machines on the non-bottleneck stages are needed to produce the same number of jobs per time unit. This allows the calculation of how many machines have to be set up on the non-bottleneck stages for each machine that is set up on the bottleneck stage, which in turn enables the incorporation of system-wide setup costs into Phase I: The associated costs of a setup on the bottleneck stage are calculated by the sum of the incurred setup costs over all production stages. Hence, the objective of Phase I is to minimize system-wide setup costs in conjunction with back-order and inventory costs of end-products.

If the bottleneck stage is the last production stage, the due dates (periods of demand) within the lot-sizing and scheduling routine can be set according to the original delivery dates. If the bottleneck stage is followed by other production stages, the due dates within Phase I must be adjusted to account for the production time on subsequent stages. This can be accomplished by a fixed lead time.

Phase II keeps the aggregation level of product families, but explicitly includes the non-bottlenecks stages. It assigns a production time and a machine to each product family unit. Thus, it solves a multi-stage scheduling and batching problem. However, the number of machines per product family and the batch-sizes are pre-determined by Phase I. The results are so-called 'machine/time slots', signalling at what time and on which machine a product family unit is to be produced. The planning process can be invoked for each period separately. The third panel of Fig. 3.2 gives an example for the second period. While the machines on each stage may not produce longer than the length of a period, the schedules for consecutive stages may overlap and cover adjacent periods of the original problem. Thus, a job may enter the system in one period and leave it in a different one.

As mentioned above, Phase II has no impact on inventory or back-order costs of end-products. However, it has an impact on inventory costs of intermediate products. Its objective is to minimize mean flow time. This implies a minimization of waiting times between stages, and thus of in-process inventory. As a side effect, the usage of intermediate buffers is also minimized.

Phase III disaggregates the product families to individual products. Each product family is considered separately. The problem is to determine how many and which machines to set up and when to produce the products of the family. All stages are considered simultaneously. Hence, Phase III solves a scheduling and batching problem for all stages on product level. The solution is generated using the machine/time slots from Phase II: An individual product unit of the respective family is assigned to each slot. The objective is to minimize intra-family setup costs while keeping the low flow time from Phase II. There is a trade-off between the two objectives. The fourth panel of

Fig. 3.2 gives an example for product family C which is assumed to consist of products $C1$ and $C2$. In this example, we have chosen to minimize flow time: We could save a setup on stage l if we exchanged the units of $C1$ on machine 2 with the units of $C2$ on machine 1. However, this would increase the flow time. Since we assume that on any production stage, all products of a family share the same setup costs, the setup cost objective may be re-written so as to minimize the number of intra-family setups (weighted by the setup costs on the respective production stage).

The overall result of the solution procedure is a detailed schedule consisting of exact time and machine assignments for each product unit in every period on all stages.

4

Phase I: Bottleneck Planning

Are you ready to battle?

In this phase, we are confronted with the following single stage lot-sizing and scheduling problem: Multiple product families have to be produced. A deterministic, discrete demand volume for every product family is given for pre-defined periods. Producing a product family consumes machine capacity, which is scarce. When changing from one product family to another, setup costs and setup times are incurred, reducing machine capacity. When a product family unit is not produced in its demand period, product family-specific inventory or back-order costs are incurred. There are several identical machines in parallel. The problem is to find an optimal production plan consisting of optimal production periods and quantities (lot-sizing) as well as machine assignments and product family sequences on the assigned machines (scheduling).

Phase I does not differentiate the individual products of a family. Thus, a product family may be considered to consist of a single product. In order to avoid a somewhat cumbersome notation, we therefore use the word 'product' when referring to a product family in the remainder of this chapter.

The outline of the chapter is as follows: Section 4.1 gives an introduction to the problem. Section 4.2 offers a literature review with respect to related lot-sizing and scheduling problems. Mixed integer programming model formulations are illustrated in Sect. 4.3. Section 4.4 presents a new solution approach. The solution approach is based on a novel model formulation given in Sect. 4.4.1, which is embedded in a solution procedure described in Sect. 4.4.2. Section 4.4.3 elaborates on using the approach in a rolling planning horizon. We report on computational results in Sect. 4.5. Section 4.6 gives a summary of the chapter.

4.1 Lot-sizing and Scheduling on Parallel Machines

Lot-sizing models can be classified as small bucket or big bucket models. Small bucket models consist of relatively short periods. They usually allow only one product or setup per period and machine. Big bucket models contain fewer but longer periods and usually have no restriction on the number of products or setups per period and machine. Big bucket models do not consider the product sequence within a period. Hence, a production schedule is not part of the solution. With small bucket models, a production schedule may directly be concluded from a solution, as the sequence of products is given by the sequence of periods and the products produced in those periods.

This general advantage of small bucket models comes at the cost of higher computation times. For the same planning horizon, the number of periods must be substantially larger than in big bucket models to yield comparable solutions. For this reason, we consider a big bucket model. In this, we will integrate the scheduling decisions. A standard big bucket model formulation—with a single machine and without back-ordering—is the Capacitated Lot-Sizing Problem (CLSP), see e.g. Billington et al. (1983) or Eppen and Martin (1987):

Parameters:

c_p^i Inventory holding costs of product p

c_p^s Setup costs of product p

C Machine capacity per period

d_{pt} Demand volume of product p in period t

P Number of products, $\bar{P} = \{1 \ldots P\}$

t_p^s Setup time of product p

t_p^u Per unit process time of product p

y_p^0 Initial inventory volume of product p at beginning of planning interval

T Number of periods, $\bar{T} = \{1 \ldots T\}$

z Big number, $z \geq \sum_{\substack{p \in \bar{P} \\ t \in \bar{T}}} d_{pt}$

Variables:

x_{pt} Production volume of product p in period t

y_{pt} Inventory volume of product p at the end of period t

γ_{pt} Binary setup variable for product p in period t

        ($\gamma_{pt} = 1$, if a setup is performed for product p in period t)

Model CLSP (Capacitated Lot-Sizing Problem)

$$\min \sum_{\substack{p \in \bar{P} \\ t \in \bar{T}}} c_p^i y_{pt} + \sum_{\substack{p \in \bar{P} \\ t \in \bar{T}}} c_p^s \gamma_{pt} \tag{4.1}$$

subject to

$$y_{p,\,t-1} + x_{pt} - d_{pt} = y_{pt} \qquad\qquad \forall\, p \in \bar{P},\, t \in \bar{T} \tag{4.2}$$

$$\sum_{p \in \bar{P}} t_p^u x_{pt} \leq C \qquad\qquad \forall\, t \in \bar{T} \tag{4.3}$$

$$x_{pt} \leq z \cdot \gamma_{pt} \qquad\qquad \forall\, p \in \bar{P},\, t \in \bar{T} \tag{4.4}$$

$$y_{p0} = y_p^0 \qquad\qquad \forall\, p \in \bar{P} \tag{4.5}$$

$$x_{pt} \geq 0,\, y_{pt} \geq 0,\, \gamma_{pt} \in \{0, 1\} \qquad\qquad \forall\, p \in \bar{P},\, t \in \bar{T} \tag{4.6}$$

The objective function (4.1) minimizes inventory and setup costs. Equations (4.2) establish the inventory flow: Demand volume of a period t must be met by inventory volume from previous periods or by a production quantity in t. Surplus volume will be stored in inventory for subsequent periods. Conditions (4.3) limit the time for processing to the machine capacity. Conditions (4.4) ensure that a setup is performed in each period that a product is produced. Equations (4.5) set the initial inventory volumes and conditions (4.6) enforce non-negative and binary variables, respectively.

The inclusion of setup times entails that the capacity must suffice not only for processing times, but also for setup times. This can be achieved by replacing constraints (4.3) with (4.7), introducing a new parameter t_p^s for the setup time of product p (e.g. Trigeiro et al. 1989):

$$\sum_{p \in \bar{P}} \left(t_p^u x_{pt} + t_p^s \gamma_{pt} \right) \leq C \qquad\qquad \forall\, t \in \bar{T} \qquad (4.7)$$

Bitran and Yanasse (1982) show that the CLSP with setup costs is an NP-hard problem, meaning that one cannot expect to find an efficient algorithm generating an optimal solution. When setup times are included, Maes et al. (1991) show that even the feasibility problem becomes NP-complete. This implies that one cannot efficiently say whether a feasible solution exists at all. Hence, as already stated by Garey and Johnson (1979), p. 3, heuristics are needed that do not guarantee an optimal solution, but find a reasonably good solution in a moderate amount of computation time. This holds true especially when real world problems are to be solved, as the computation time increases with the problem size.

The approach presented in this study extends the above problem formulation found in the literature in, mainly, three ways:

Firstly, because capacity is scarce, it might be useful to produce a product volume in a period other than its demand period to save setup time and costs. Traditional lot-sizing models allow a product to be produced in a period before its delivery to the customer. As a consequence, inventory costs occur. In our case, it is also possible that the product cannot be delivered on time. It is then back-ordered and associated **back-order** costs are incurred for every unit and period of the delay. Figure 4.1 illustrates an example of a single product with a demand in period 3. Production takes place in periods 1, 2, 3 and 4. Without the possibility to back-order product units, no feasible plan would exist because of the capacity constraint.

Secondly, it is possible to carry over a setup state from one period to another. If, on a machine, the last product of a period and the first of a subsequent period are the same, no setup has to be performed. In standard big bucket lot-sizing models, a setup has to be performed in each period in which a product is produced. If a model leaves **setup carry-over** unconsidered, it might not be able to find feasible solutions, because too much capacity is consumed by setup times that are not needed in reality. In Fig. 4.2, the setup for product a in period 2 can be saved because a setup carry-over from period 1 can be used. Thus, the additional capacity can be employed to set up the machine and produce another product b. Haase (1998) points out that solutions become significantly different when setup carry-over is considered.

The inclusion of setup carry-over entails a partial order of the products in a period, leading to a model that incorporates characteristics of a small bucket

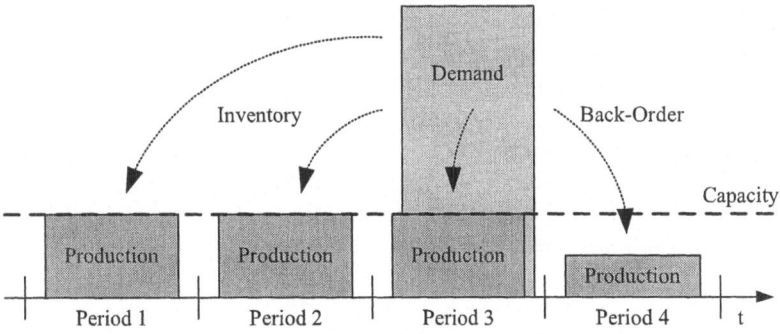

Fig. 4.1. Inventory holding and back-ordering of products

model into a big bucket model. It also intertwines lot-sizing and scheduling decisions: The lot-sizing phase can only determine the production volume of a period when the machine capacity is known. The machine capacity depends on the amount of time needed for setups, which in turn depends on the job sequence and the setup carry-over, and therefore on the scheduling phase. On the other hand, the job sequence cannot be generated without knowing which products are produced in a period, a result of the lot-sizing phase. Ultimately, both phases have to be solved in an integrated way.

Thirdly, we consider **parallel machines**. In the literature, capacity constraints are generally modeled by an aggregate facility, i.e. by a single (aggregated) machine. Such solution procedures might suggest to lump together demands of subsequent periods to produce the whole volume in a single period. With only one machine and sufficient capacity, this method would save setup time and costs. In our case, producing more units in a period might lead to the result that more parallel machines have to be set up for the product. Together with the setup carry-over, that approach would lead to higher setup times and costs. A better solution would be to produce in the demand periods, saving back-order and/or inventory costs. Fewer machines in parallel would be

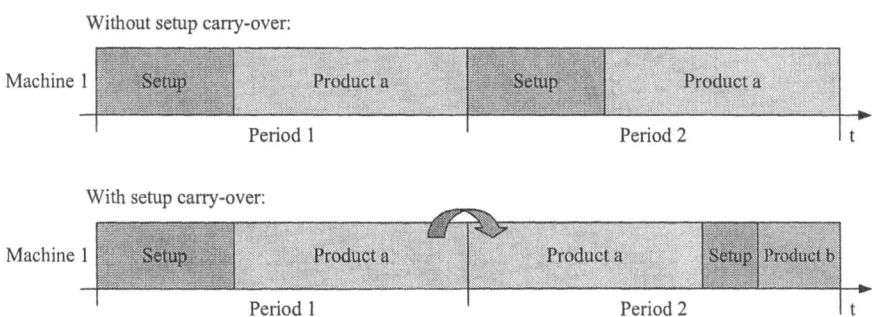

Fig. 4.2. Setup carry-over

needed, and the setup could be carried over between the periods. Figure 4.3 illustrates this with an example of a single product that has an equally distributed demand over three consecutive periods. In solution 1, six machines are set up for the product. Furthermore, inventory holding and back-order costs are incurred as some processing does not take place in the demand periods. In solution 2, all production takes place in the demand periods, avoiding inventory holding and back-order costs. Only two machines in parallel are set up and their setup state is carried over from period 1 to 2 and 3.

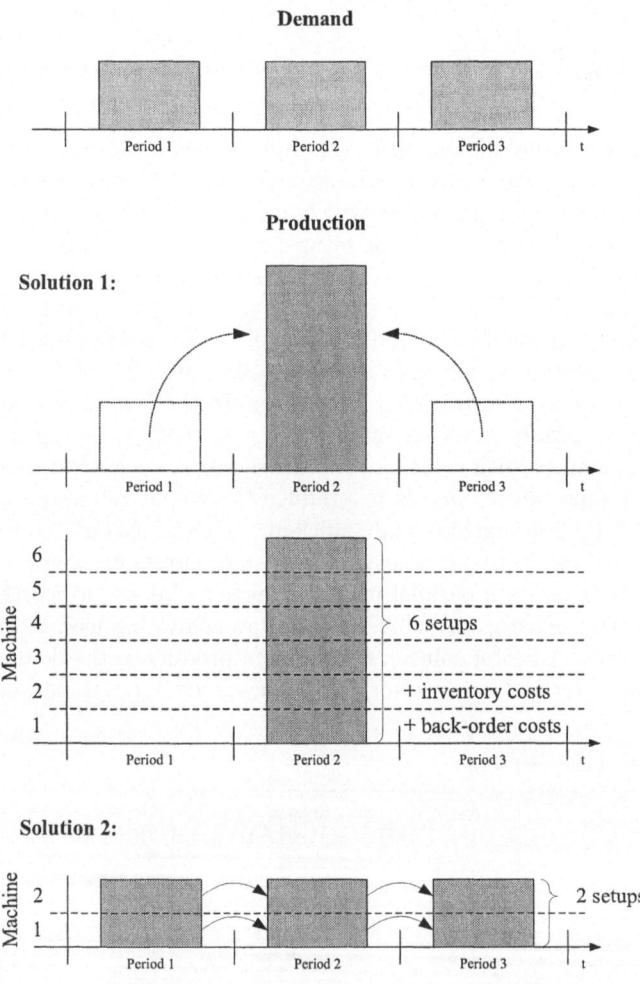

Fig. 4.3. Parallel machines and setup carry-over constitute a different lot-sizing problem

4.2 Literature Review

Lot-sizing problems have been addressed extensively in the literature. An overview of (mainly) small bucket models and solution procedures can be found in Drexl and Kimms (1997). Research integrating lot-sizing and scheduling decisions in small and big bucket models is reviewed by Meyr (1999).

In this section, we focus on capacitated big bucket models and their solution procedures. Various formulations of the big bucket CLSP and solution approaches including extensions to multi-level product structures are well known (e.g. Eppen and Martin 1987; Trigeiro et al. 1989; Millar and Yang 1993; Tempelmeier and Derstroff 1996; Stadtler 1996; Katok et al. 1998). Reviews and taxonomies of such models and solution procedures are given by Bahl et al. (1987), Maes and Van Wassenhove (1988), Kuik et al. (1994), Helber (1994) and Derstroff (1995).

Maes and Van Wassenhove (1988) differentiate between so-called 'common-sense' heuristics and 'mathematical programming' approaches to solve lot-sizing problems. The latter ones are based on standard mathematical optimization procedures like Lagrangian relaxation, Branch&Bound or Linear Programming (see Thizy and Van Wassenhove 1985; Gelders et al. 1986; Maes et al. 1991 for examples). The methods seek the global optimum and are usually truncated in some way to reduce the computational effort. On the other hand, common-sense heuristics use some insights into the lot-sizing problem itself and exploit its special structure. Examples are 'period-by-period' and 'improvement' heuristics. While improvement heuristics (e.g. Dogramaci et al. 1981) start from a possibly infeasible solution and try to find a good, feasible one by shifting production volumes to different periods, period-by-period heuristics determine the production volumes of each period separately, in a (forward or backward) chronological manner. Usually, they employ priority-rules to determine the production volumes. One problem of period-by-period heuristics is the anticipation of capacity bottlenecks in future periods, especially because back-orders are not allowed in the standard CLSP. Two subclasses may be distinguished by the way they try to ensure feasibility: Some methods anticipate future bottlenecks a-priori during the forward movement ('strict-forward' or 'single-pass' heuristics, e.g. Dixon and Silver 1981), others employ a feedback mechanism to shift excess demand to leftover capacity in previous periods when they have detected a lack of capacity in the actual period (see Lambrecht and Vanderveken 1979 for an example).

There are distinct advantages and disadvantages of both the mathematical programming and the common-sense approaches. Since the latter use the special structure of the specific lot-sizing problem at hand, they are not easily adaptable to different problem types. For example, a common-sense heuristic might not be able to include a constraint on the aggregate inventory level. In contrast, mathematical programming approaches are quite general. An aggregate inventory restriction could easily be included using a linear constraint in the model formulation. Mathematical programming approaches also tend to find better solutions (see Maes and Van Wassenhove 1988), but at the cost of a much higher computational effort.

Several common sense heuristics are compared by Maes and Van Wassenhove (1986a, 1986b). They find out that a period-by-period heuristic with an a-priori anticipation of future bottlenecks yields better solutions than one with a feedback mechanism. Nevertheless, there are cases in which the opposite is true. Moreover, it is left open if this is a structural disadvantage or just a matter of the specific feedback mechanism used in the example.

4.2.1 Back-Order Literature

Despite its importance in practical settings, only few researchers have addressed capacitated lot-sizing problems with back-ordering. Smith-Daniels and Smith-Daniels (1986) present a single machine mixed integer programming model with back-orders that is a combination of a small bucket and a big bucket approach. In each period, only one product family can be produced. A family may consist of various items. While there are sequence-dependent setup times between items, no setup time is incurred between families as family setups are supposed to take place between periods. On the other hand, there are sequence-independent family setup costs but no item-to-item setup costs. As usual in small bucket models, a family can carry over its setup to a subsequent period. A simplified version of the model including 4 items and 5 periods is solved using standard procedures. In Smith-Daniels and Ritzman (1988), the model is extended to the multi-level case. In this version, more than one family may be produced per period, making it a big bucket model. Now, a sequence-dependent setup time is incurred for family changeovers, but item-to-item setup times do not accrue. Setup costs are not considered at all. The approach automatically sequences the families within a period and is able to carry over a family setup from one period to another. An example instance consisting of 3 families with 2 items each, 3 periods and 2 stages is solved using standard procedures.

Pochet and Wolsey (1988) tackle the single machine CLSP with back-orders using a shortest-path and an alternative plant location formulation. The first one is solved by standard mixed integer programming procedures, the second by a cutting plane algorithm. Setup times and setup carry-over are not considered. They solve problems of up to 100 products and 8 periods.

Millar and Yang (1994) extend an approach from Millar and Yang (1993) and present two algorithms for a single machine lot-sizing problem with back-orders. One algorithm is based on Lagrangian decomposition, the other on Lagrangian relaxation, both neither cover setup times nor setup carry-over. Their test problems consist of 5 products and 6 periods.

Cheng et al. (2001) consider the analogies between the CLSP with back-orders and the traditional fixed charge transportation problem. Their solution procedure is three-phased. After heuristically determining setups for some products in certain periods, they use a procedure similar to Vogel's approximation method for the transportation problem to create an initial solution. Afterwards, they employ a variation of the standard primal transportation simplex to improve the solution. Despite covering only a single machine, their model takes different kinds of capacity sources into consideration. Only a single setup has to be performed when one or more of the capacity sources are used in a period, but the production costs differ between capacity sources. This allows the modeling of regular time and overtime. Setup times and setup carry-over are not considered. Cheng et al. solve problems with 3 products and 12 periods.

4.2.2 Setup Carry-Over Literature

Dillenberger et al. (1993) cover the CLSP with setup carry-over but without back-ordering. Their research is motivated by a practical production planning problem and is slightly different from the one usually found in the lot-sizing literature: Production of a product must always take place in its demand period. If the capacity does not allow the complete demand to be met, the surplus volume is not produced at all and the demand is lost. The approach covers setup times and parallel machines. Dillenberger et al. (1994) extend the model to cover different kinds of resources such as energy or components. Some resources are storable, meaning if their capacity is not consumed in a period, it may be used in later periods. Products are grouped to product families. Changing between families incurs major setups, while changing products within a family incurs minor setups, leading to partially sequence-dependent setups. Both major and minor setups may incur setup times and costs. The solution procedure consists of a partial Branch&Bound scheme that iterates on a period-by-period basis. In each iteration, only setup variables relating to the actual period are fixed to binary values, while the binary restrictions for later periods are relaxed. The authors solve practical problems with up to 40 products, 6 periods and 15 machines.

Haase (1994) establishes the name CLSPL for a CLSP with setup carry-over (Capacitated Lot-Sizing Problem with Linked lot-sizes). He solves a standard CLSPL with a single machine and without back-ordering. Setup times are not considered. His procedure moves backwards from the last to the first period and schedules setups and production volumes using a randomized regret measure. He points out that, from a conceptual point of view, a backward-oriented procedure is superior when solving a standard CLSPL: A forward-oriented procedure must decide how many units to produce for future demands, depending on complicated and time-consuming estimations of future capacity availabilities. Without back-orders, a backward-oriented procedure can always focus on the actual period as remaining unmet demands of later periods are already known. Clearly, with back-orders allowed, this observation is obsolete, as a backward-oriented procedure would have to decide how many units to back-order just as a forward-oriented procedure must decide how many units to store in inventory. Haase solves well known problems from Thizy and Van Wassenhove (1985) including 8 products and 8 periods. Later, Haase (1998) reduces the model formulation so that a setup state cannot be preserved over more than one period boundary. The resulting model is much easier to solve and the solution quality is similar—given his test problems covering a single machine. He solves problems with 50 products and 8 periods, 20 products and 20 periods as well as 8 products and 50 periods.

Gopalakrishnan et al. (1995) present a new mixed integer programming model for a variation of the CLSPL covering parallel machines but no back-ordering. Setup times and costs are included, but are product-independent—i.e. they are the same for all products. The formulation makes extensive use of binary variables when compared with other formulations. They solve a practical problem covering 12 periods, 3 machines and 2 product families using standard procedures. Gopalakrishnan (2000) considers a slightly different problem. In this study, setup times and costs are product-dependent. On the other hand, the new model formulation includes only a single machine. An example instance with 2 products and 3 periods is solved to optimality using standard procedures. Gopalakrishnan et al. (2001) develop a Tabu Search algorithm for this problem and solve problems with 30 products and 20 periods.

Also Sox and Gao (1999) develop a new model for the CLSPL. Their formulation covers a single machine, no back-ordering and no setup times. It differs from the approach by Dillenberger et al. (1994) and Haase (1994), as it does not contain a binary variable indicating that a product is the only product produced in a specific period (on a certain machine). A shortest-path reformulation of the model is solved optimally for smaller test instances covering 8 items and 8 periods. For larger problems, the setup carry-over is restricted to one subsequent period per product as suggested by Haase (1998). Using a Lagrangian decomposition heuristic, instances with up to 100 products and 10 periods are solved to near-optimality in less than one minute computation time on a Sparcstation 20. Gao (2000) proposes a simple forward period-by-period heuristic for the same restricted problem. In one test, the instances cover 50 products and 8 periods, in another test 20 products and 20 periods. The tests indicate that the new heuristic requires shorter computation times but leads to higher costs.

Sürie and Stadtler (2003) use a simple plant location reformulation of the CLSPL as introduced by Haase (1994). Their approach covers setup times and multiple machines in a multi-level production environment, but the product/machine assignment is unique, meaning that the machines cannot be used in parallel. Back-ordering is not allowed. They employ a Branch&Cut and a Cut&Branch algorithm, which add valid inequalities to the model and solve the test instances described by Gopalakrishnan et al. (2001). For larger instances of the test, they suggest a time-oriented decomposition approach that moves a lot-sizing time-window through the planning periods. In each of its iterations, the problem is solved for all periods, but the complete set of constraints is only enforced for the periods of the time-window. Periods before the time-window have been planned in an earlier iteration and their binary variables are fixed. In the time interval following the lot-sizing window, capacity utilization is only estimated by neglecting setups.

Lot-sizing models with setup carry-over determine the last product in each period. Thus, as described above, they establish a partial order of the products. Some researches have developed models and procedures that construct a total sequence of all products in big bucket models. The last product of the sequence can carry over its setup to the next period. So far, these approaches also deal with setup carry-over.

Aras and Swanson (1982) develop a backward-oriented period-by-period loading procedure for a single machine problem with sequence-independent setup times. As setup costs are not considered, the objective is to minimize total holding costs. Holding costs are taken into account even within the period of production. Therefore, a total sequence of all products is generated and the machine can carry over its setup state to the next period. Back-orders are not allowed. They solve a practical problem covering 38 products and 26 periods.

Selen and Heuts (1990) discuss a lot-sizing and scheduling problem in a chemical manufacturing environment where production lots must consist of an integer number of pre-defined quantities (batches). The problem covers a single machine. In addition to standard problems, the inventory capacity is limited. Setup times are sequence-dependent and setup carry-overs are possible. Back-orders are not allowed. The proposed solution procedure starts with a feasible production plan based on a lot-for-lot solution. It shifts complete production lots to earlier periods, with the target period moving forward in a period-by-period manner. A sample problem with 20 products and 10 periods is solved. Heuts et al. (1992) compare two modified versions of the described heuristic in a rolling horizon setting. One modification moves the source period for a production shift instead of the target period. In the other modification, all periods are eligible as source and target period at once. Both procedures lead to similar results in different computational experiments with 15 products and 34 periods (out of which 10 are considered in each run due to the rolling horizon).

Haase (1996) formulates a big bucket lot-sizing model creating a total order of all products. It contains a single machine. Setup times and back-orders are not considered. He uses a variation of the backward-oriented, randomized regret based heuristic described by Haase (1994) to solve the problem. The test instances cover up to 9 products and 8 periods or 4 products and 26 periods. Haase and Kimms (2000) present a new model that incorporates sequence-dependent setup times. For each period, the model has to select a sequence of products from an a-priori given set of such sequences and to determine the production volumes. In its essence, the solution procedure is a Branch&Bound method. It moves backwards from the last period to the first one, selecting one of the given sequences in each iteration. The size of solvable instances ranges from 3 products and 15 periods to 10 products and 3 periods.

Grünert (1998) presents a multi-machine model for the multi-level case. However, a unique assignment of products to machines must be given and hence the machines cannot be used in parallel. The model contains sequence-dependent setup times and an option to use overtime. Back-orders are not allowed. Tabu Search, a metaheuristic, is employed in conjunction with La-grangian decomposition. The generation of neighboring solutions during the Tabu Search procedure is guided by the solution of the Lagrangian problem. When the latter suggests a setup for a product in a certain period and the setup is not tabu, it will be included in the new candidate solution. Problems with up to 300 products and 10 periods or 10 products and 30 periods are tested, all with one machine per production level. The number of production levels in the test problems is a result of the stochastic instance generator and can therefore not be ascertained. Nevertheless, Grünert mentions that com-putation times become prohibitive when more than 10 products have to be planned on a single machine (p. 259).

Fleischmann and Meyr (1997) develop a model that can be viewed as both a small bucket and a big bucket model: Each 'macro-period' consists of (pos-sibly empty) 'micro-periods' of variable length. While the number of products per micro-period is restricted to one, the number of products per macro-period is potentially unbounded. The model contains a single machine and establishes a total order of all products. Setup times and back-orders are not considered. They suggest a solution method based on Threshold Accepting, a metaheuristic. Starting from an infeasible solution, the method generates new setup sequences by performing neighborhood operations such as insert-ing a setup for a certain product, exchanging two setups or deleting one. Each setup sequence is evaluated using a heuristic procedure that determines the production and inventory volumes. Computational tests are performed us-ing the instances from Haase (1996). Meyr (2000) states that once the setup sequences are fixed, the remaining problem of determining production and inventory volumes becomes a linear programming (LP) problem. He drops the heuristic sub-procedure and replaces it with an efficient network flow al-gorithm to evaluate the setup sequences optimally. The algorithm exploits the fact that the setup sequences differ only marginally and thus, a solution may be found using re-optimization. Besides Threshold Accepting, he also uses Simulated Annealing as a coordinating procedure and extends the model to sequence-dependent setup times. In addition to a comparison with the re-sults of Haase (1996), he solves problems with setup times covering up to 18 products and 8 periods.

Laguna (1999) introduces another model. It also contains a single machine and sequence-dependent setup times, and extends the problem by an overtime option. Again, back-orders are not allowed. A solution approach similar to the one by Meyr (2000) is suggested. The method starts with an LP-relaxation of the model and a following Traveling Salesman algorithm to order the products of each period. Production and inventory volumes of an initial solution are determined by solving another mixed integer program. Afterwards, a Tabu

Search procedure is invoked to improve the solution by neighborhood operations. Each solution is evaluated using the Traveling Salesman procedure to re-sequence the products and the mixed integer program to find production and inventory volumes. Both these sub-problems are solved to optimality for each candidate using standard procedures. Test instances cover 3 products and 12 periods or 5 products and 3 periods.

4.2.3 Parallel Machines Literature

Besides the articles by Dillenberger et al. (1993), Dillenberger et al. (1994) and Gopalakrishnan et al. (1995), which have already been mentioned above, little research is available that integrates parallel machines into big bucket lot-sizing models.

Diaby et al. (1992) include parallel resources in the CLSP. Their formulation contains setup times, but no setup carry-over and back-orders. Only a single setup has to be performed whenever a product is produced in a period, no matter how many resources are being used. While this assumption is valid when parallel resources are employed to represent regular time and overtime or similar concepts, it is not suited for the parallel machines concept used here. Using a Lagrangian relaxation scheme with subgradient optimization, they solve instances with up to 5 000 products and 30 periods.

Derstroff (1995) solves a multi-level parallel machine CLSP, also using a Lagrangian relaxation procedure. In a first step, the model is solved with relaxed capacity and inventory flow constraints. In a second step, the Lagrangian multipliers are updated, and in a third step, a feasible solution is created by shifting production volumes to different periods and machines. While setup times are covered, setup carry-over and back-ordering are not considered. The test instances include 20 products on 5 production levels, 16 periods and 6 machines, of which up to two may be used in parallel.

Hindi (1995) considers a parallel machine lot-sizing problem *without* setups. Back-orders are not allowed. The problem is reformulated as a capacitated transshipment model and solved using a primal and a dual network simplex algorithm. Instances with up to 200 items, 24 periods and 2 machines are solved on a Sun ELC workstation.

Özdamar and Birbil (1998) tackle a parallel machine lot-sizing problem with setup times but without setup costs. In addition to regular capacity, each machine is allowed to use a certain amount of overtime capacity at respective costs. Back-ordering and setup carry-over are not allowed. The parallel machines concept differs from the one described here, as lot-splitting is not allowed—i.e. the complete per-period production volume of a product must be produced on a single machine. In other words, only one of the parallel machines may be used to produce a specific product per period. Three similar hybrid heuristics are developed to solve the problem. In each of them, a Genetic Algorithm operates mainly in the infeasible region of the solution space. An interweaved Tabu Search and Simulated Annealing algorithm is coupled to improve single solutions of a population and make them feasible. Sample problems with up to 20 products, 6 periods and 5 machines are solved. Özdamar and Barbarosoglu (1999) extend the problem to multiple production stages. They also include setup times and the possibility to back-order end-products. Lot splitting remains prohibited. A stand-alone Simulated Annealing procedure is compared with two hybrid heuristics combining Simulated Annealing with a Genetic Algorithm (in a manner similar to the above hybrid heuristics) and a Lagrangian relaxation approach, respectively. The latter provides better results with respect to solution quality and computation time for test problems with up to 20 products, 6 periods and 5 machines on each of the 4 production stages.

Kang et al. (1999) present a model and solution procedure tailor-made for the so-called 'CHES problems' described by Baker and Muckstadt (1989). The model establishes a total order of all products on each machine, allowing the setup state to be carried over to a subsequent period. Setup times and back-orders are not included. The problems are based on real-life cases and cover up to 21 products, 3 periods and 10 machines. However, only two of the five problems cover more than one period, and those include only one or two machines. Unlike usual lot-sizing problems, the objective is to maximize the profit contribution: Products are sold at market prices and the given demand may be exceeded. Kang et al. (1999) extend the test-bed by instances covering 6 products, 9 periods and 1 or 2 machines. The problems are solved by a truncated Branch&Bound procedure that incorporates a Column Generation approach for the LP-relaxations at each node. Afterwards, the solution is improved using a local search method.

Meyr (2002) extends the algorithm described by Meyr (2000) to work with parallel machines. Again, he uses a Threshold Accepting and a Simulated Annealing approach to fix the setup sequence and a network flow algorithm to determine the production and inventory volumes. However, compared with the single machine case, the network flow algorithm becomes more complicated and time-consuming when non-identical machines are considered. Meyr evaluates the approach with test instances covering 15–19 products, 8 periods and 2 machines. A modified version of the heuristic is also compared with the one of Kang et al. (1999). Meyr (1999) suggests a decomposition approach for larger problems, based on the above algorithms. The procedure is explained using a real-world case, but no numerical tests are given.

Table 4.1 summarizes the literature review. To our knowledge, there is no literature incorporating general parallel machines in lot-sizing approaches with back-orders.

Table 4.1. Literature review of capacitated big bucket lot-sizing models with respect to back-ordering, setup times, setup carry-over and parallel machines.
x: covered in the reference; (x) and ((x)): partly covered in the reference, see text.

Reference	Back-orders	Setup times	Setup carry-over	Parallel machines
Smith-Daniels and Smith-Daniels (1986)	x	(x)	x	
Smith-Daniels and Ritzman (1988)	x	x	x	
Pochet and Wolsey (1988)	x			
Millar and Yang (1994)	x			
Cheng et al. (2001)	x			
Dillenberger et al. (1993)		x	x	x
Dillenberger et al. (1994)		x	x	x
Haase (1994)			x	
Haase (1998)			x	
Gopalakrishnan et al. (1995)		(x)	x	x
Gopalakrishnan (2000)		x	x	
Gopalakrishnan et al. (2001)		x	x	
Sox and Gao (1999)			x	
Gao (2000)			x	
Sürie and Stadtler (2003)		x	x	
Aras and Swanson (1982)		x	x	
Selen and Heuts (1990)		x	x	
Heuts et al. (1992)		x	x	
Haase (1996)			x	
Haase and Kimms (2000)		x	x	
Grünert (1998)		x	x	
Fleischmann and Meyr (1997)			x	
Meyr (2000)		x	x	
Laguna (1999)		x	x	
Diaby et al. (1992)		x		((x))
Derstroff (1995)		x		x
Hindi (1995)				x
Özdamar and Birbil (1998)		x		(x)
Özdamar and Barbarosoglu (1999)	x	x		(x)
Kang et al. (1999)			x	x
Meyr (1999)		x	x	x
Meyr (2002)		x	x	x

4.3 Standard Models with Extensions

The lot-sizing and scheduling problem at hand may be formulated as a mixed integer programming (MIP) model. A MIP model represents the problem in a mathematical way, precisely defining its characteristics. In this section, we present model formulations covering back-orders, parallel machines, setup times and setup carry-over. They are based on a standard big bucket formulation, the Capacitated Lot-Sizing Problem (CLSP) as presented above.

4.3.1 Capacitated Lot-Sizing Problem with Linked Lot-Sizes (CLSPL)

The CLSPL (Capacitated Lot-Sizing Problem with Linked lot-sizes) is an extension of the CLSP regarding setup carry-over. We present it including setup times, based on a formulation by Sox and Gao (1999). In addition to the notation for the CLSP, a new set of parameters and a new set of variables are used:

Parameters:

ζ_p^0 Initial setup state of product p at beginning of planning interval ($\zeta_p^0 = 1$, if the machine is initially set up for product p)

Variables:

ζ_{pt} Binary linking variable for product p in period t ($\zeta_{pt} = 1$, if the setup state for product p is carried over from period t to $t+1$)

Model CLSPL (CLSP with Linked lot-sizes (setup carry-over))

$$\min \sum_{\substack{p \in \bar{P} \\ t \in \bar{T}}} c_p^i y_{pt} + \sum_{\substack{p \in \bar{P} \\ t \in \bar{T}}} c_p^s \gamma_{pt} \qquad (4.8)$$

subject to

$$y_{p,\,t-1} + x_{pt} - d_{pt} = y_{pt} \qquad \forall\, p \in \bar{P},\, t \in \bar{T} \qquad (4.9)$$

$$\sum_{p \in \bar{P}} \left(t_p^u x_{pt} + t_p^s \gamma_{pt} \right) \leq C \qquad \forall\, t \in \bar{T} \qquad (4.10)$$

$$x_{pt} \leq z \left(\gamma_{pt} + \zeta_{p,\,t-1} \right) \qquad \forall\, p \in \bar{P},\, t \in \bar{T} \qquad (4.11)$$

$$\sum_{p \in \bar{P}} \zeta_{pt} = 1 \qquad \forall\, t \in \bar{T} \qquad (4.12)$$

$$\zeta_{pt} - \gamma_{pt} - \zeta_{p,\,t-1} \leq 0 \qquad \forall\, p \in \bar{P},\, t \in \bar{T} \qquad (4.13)$$

$$\zeta_{pt} + \zeta_{p,\,t-1} - \gamma_{pt} + \gamma_{qt} \leq 2 \qquad\qquad \forall\, p, q \in \bar{P},\ q \neq p,\ t \in \bar{T} \quad (4.14)$$

$$y_{p0} = y_p^0,\ \zeta_{p0} = \zeta_p^0 \qquad\qquad\qquad\qquad \forall\, p \in \bar{P} \qquad\qquad\quad (4.15)$$

$$x_{pt} \geq 0,\ y_{pt} \geq 0,\ \gamma_{pt} \in \{0, 1\},\ \zeta_{pt} \in \{0, 1\} \quad \forall\, p \in \bar{P},\ t \in \bar{T} \qquad (4.16)$$

The differences to the CLSP are revealed in constraints (4.11) to (4.14). Conditions (4.11) ensure that item p can only be produced in a period t if the machine is set up for it. With setup carry-over, this can be achieved in two ways—by either carrying over a setup state from period $t-1$ (in which case $\zeta_{p,\,t-1} = 1$), or by performing a setup in period t ($\gamma_{pt} = 1$). Conditions (4.12) to (4.14) handle the correct implementation of the setup carry-over. Conditions (4.12) imply that the setup state can be carried over for one product only. Conditions (4.13) state that the machine must be set up for a certain product in order to carry over the setup state for this product to the next period. That is, if the machine carries over a setup state for product p from period t to $t+1$ ($\zeta_{pt} = 1$), the machine must have been set up for this product in period t ($\gamma_{pt} = 1$), or the setup state had been carried over from the previous period ($\zeta_{p,\,t-1} = 1$). Conditions (4.14) deal with the following situation: If the machine carries over a setup state for item p from period $t-1$ to t and also from t to $t+1$ ($\zeta_{p,\,t-1} = \zeta_{pt} = 1$) and, in period t, a setup is performed for another product q ($\gamma_{qt} = 1$), then we have to re-set up the machine to p in period t ($\gamma_{pt} = 1$). Finally, compared with their CLSP counterparts (4.5) and (4.6), equations (4.15) include the initialization of the setup state and conditions (4.16) are augmented to enforce binary setup carry-over variables.

4.3.2 CLSPL with Back-Orders and Parallel Machines (CLSPL-BOPM)

The CLSPL may be extended to cover back-orders and parallel machines. For parallel machines, the production, setup and setup carry-over variables as well as the initial setup state parameter receive an additional machine index. For back-orders, additional back-order volume variables as well as associated parameters are included. The following complete notation is used:

Parameters:

b_p^0	Initial back-order volume of product p at beginning of planning interval
c_p^b	Back-order costs of product p
c_p^i	Inventory holding costs of product p
c_p^s	(System-wide) setup costs of product p
C	Capacity of a parallel machine per period
d_{pt}	Demand volume of product p in period t
M	Number of parallel machines, $\bar{M} = \{1 \ldots M\}$

P	Number of products, $\bar{P} = \{1 \ldots P\}$
t_p^s	Setup time of product p
t_p^u	Per unit process time of product p
y_p^0	Initial inventory volume of product p at beginning of planning interval
T	Number of periods, $\bar{T} = \{1 \ldots T\}$
z	Big number, $z \geq \sum_{\substack{p \in \bar{P} \\ t \in \bar{T}}} d_{pt}$
ζ_{pm}^0	Initial setup state of product p on machine m at beginning of planning interval ($\zeta_{pm}^0 = 1$, if machine m is initially set up for product p)

Variables:

b_{pt}	Back-order volume of product p at the end of period t
x_{ptm}	Production volume of product p in period t on machine m
y_{pt}	Inventory volume of product p at the end of period t
γ_{ptm}	Binary setup variable for product p in period t on machine m ($\gamma_{ptm} = 1$, if a setup is performed for product p in period t on machine m)
ζ_{ptm}	Binary linking variable for product p in period t on machine m ($\zeta_{ptm} = 1$, if the setup state for product p on machine m is carried over from period t to $t+1$)

Model CLSPL-BOPM (CLSPL with Back-Orders and Parallel Machines)

$$\min \sum_{\substack{p \in \bar{P} \\ t \in \bar{T}}} c_p^i y_{pt} + \sum_{\substack{p \in \bar{P} \\ t \in \bar{T}}} c_p^b b_{pt} + \sum_{\substack{p \in \bar{P} \\ t \in \bar{T} \\ m \in \bar{M}}} c_p^s \gamma_{ptm} \tag{4.17}$$

subject to

$$y_{p,\,t-1} - b_{p,\,t-1} + \sum_{m \in \bar{M}} x_{ptm} - d_{pt} \qquad \forall\, p \in \bar{P},\, t \in \bar{T} \tag{4.18}$$
$$= y_{pt} - b_{pt}$$

$$\sum_{p \in \bar{P}} \left(t_p^u x_{ptm} + t_p^s \gamma_{ptm} \right) \leq C \qquad \forall\, t \in \bar{T},\, m \in \bar{M} \tag{4.19}$$

$$x_{ptm} \leq z \left(\gamma_{ptm} + \zeta_{p,\,t-1,\,m} \right) \qquad \forall\, p \in \bar{P},\, t \in \bar{T},\, m \in \bar{M} \tag{4.20}$$

$$\sum_{p \in \bar{P}} \zeta_{ptm} = 1 \qquad \forall\, t \in \bar{T},\, m \in \bar{M} \tag{4.21}$$

$$\zeta_{ptm} - \gamma_{ptm} - \zeta_{p,t-1,m} \leq 0 \qquad \forall\, p \in \bar{P},\, t \in \bar{T},\, m \in \bar{M} \qquad (4.22)$$

$$\zeta_{ptm} + \zeta_{p,t-1,m} - \gamma_{ptm} + \gamma_{qtm} \leq 2 \qquad \begin{aligned} &\forall\, p,q \in \bar{P},\, q \neq p, \\ &t \in \bar{T},\, m \in \bar{M} \end{aligned} \qquad (4.23)$$

$$b_{pT} = 0 \qquad \forall\, p \in \bar{P} \qquad (4.24)$$

$$b_{p0} = b_p^0,\; y_{p0} = y_p^0,\; \zeta_{p0m} = \zeta_{pm}^0 \qquad \forall\, p \in \bar{P},\, m \in \bar{M} \qquad (4.25)$$

$$b_{pt} \geq 0,\; x_{ptm} \geq 0,\; y_{pt} \geq 0, \qquad \forall\, p \in \bar{P},\, t \in \bar{T},\, m \in \bar{M} \qquad (4.26)$$
$$\gamma_{ptm} \in \{0,1\},\, \zeta_{ptm} \in \{0,1\}$$

The objective function (4.17) minimizes inventory, back-order and setup costs. Equations (4.18) are ordinary inventory flow conditions, augmented by back-order variables: The demand volume of a period t plus the back-order volume of previous periods must be met by inventory volume from previous periods or by a production quantity in t. If the demand volume cannot be met, it will be back-ordered to the next period while surplus volume will be stored in inventory. Conditions (4.19) to (4.23) are analogous to the CLSPL formulation, extended by a machine index. Equations (4.24) ensure that the whole demand volume is produced until the end of the planning horizon. Equations (4.25) include setting the initial back-order volumes, and conditions (4.26) incorporate non-negative back-order variables.

4.3.3 Capacitated Lot-Sizing Problem with Sequence-Dependent Setups, Back-Orders and Parallel Machines (CLSD-BOPM)

The CLSPL determines the last product per period and thus generates a partial order of the products produced on each machine. The CLSD (Capacitated Lot-sizing problem with Sequence-Dependent setups) goes further by establishing a complete order of all products per machine. As its name implies, the purpose of this is to be able to include sequence-dependent setup costs and times. However, sequence-dependent setup costs or times are not covered in this study, even though the following solution procedure calculates a complete sequence of all products per machine. Thus, the CLSD is presented with sequence-independent setup costs and times. Sequence-dependency may easily be incorporated by replacing the single-indexed setup cost and time parameters c_p^s and t_p^s with double-indexed parameters c_{qp}^s and t_{qp}^s, which denote the setup costs and times from product q to product p for each combination separately. The following formulation of the CLSD with back-orders and parallel machines is based on a formulation by Haase (1996), who presents it for a single machine without setup times and back-orders. Some additional notation is used:

Variables:

γ_{qptm} Binary setup variable indicating a setup from product q to product p in period t on machine m ($\gamma_{qptm} = 1$, if a setup is performed from product q to product p in period t on machine m)

π_{ptm} Sequencing variable indicating the ordinal position of product p on machine m in period t. The larger π_{ptm}, the later product p is scheduled on m in t

Model CLSD-BOPM (CLSD with Back-Orders and Parallel Machines)

$$\min \sum_{\substack{p \in \bar{P} \\ t \in \bar{T}}} c_p^i y_{pt} + \sum_{\substack{p \in \bar{P} \\ t \in \bar{T}}} c_p^b b_{pt} + \sum_{\substack{p, q \in \bar{P} \\ t \in \bar{T} \\ m \in \bar{M}}} c_p^s \gamma_{qptm} \tag{4.27}$$

subject to

$$y_{p,t-1} - b_{p,t-1} + \sum_{m \in \bar{M}} x_{ptm} - d_{pt} \qquad \forall\, p \in \bar{P},\, t \in \bar{T} \tag{4.28}$$
$$= y_{pt} - b_{pt}$$

$$\sum_{p \in \bar{P}} \left(t_p^u x_{ptm} + \sum_{q \in \bar{P}} t_p^s \gamma_{qptm} \right) \leq C \qquad \forall\, t \in \bar{T},\, m \in \bar{M} \tag{4.29}$$

$$x_{ptm} \leq z \left(\zeta_{p,t-1,m} + \sum_{q \in \bar{P}} \gamma_{qptm} \right) \qquad \forall\, p \in \bar{P},\, t \in \bar{T},\, m \in \bar{M} \tag{4.30}$$

$$\sum_{p \in \bar{P}} \zeta_{ptm} = 1 \qquad \forall\, t \in \bar{T},\, m \in \bar{M} \tag{4.31}$$

$$\sum_{q \in \bar{P}} \gamma_{qptm} + \zeta_{p,t-1,m} = \sum_{r \in \bar{P}} \gamma_{prtm} + \zeta_{ptm} \quad \forall\, p \in \bar{P},\, t \in \bar{T},\, m \in \bar{M} \tag{4.32}$$

$$\pi_{ptm} \geq \pi_{qtm} + 1 - P\left(1 - \gamma_{qptm}\right) \qquad \forall\, p, q \in \bar{P},\, t \in \bar{T},\, m \in \bar{M} \tag{4.33}$$

$$b_{pT} = 0 \qquad \forall\, p \in \bar{P} \tag{4.34}$$

$$b_{p0} = b_p^0,\; y_{p0} = y_p^0,\; \zeta_{p0m} = \zeta_{pm}^0 \qquad \forall\, p \in \bar{P},\, m \in \bar{M} \tag{4.35}$$

$$b_{pt} \geq 0,\; x_{ptm} \geq 0,\; y_{pt} \geq 0,\; \pi_{ptm} \geq 0,$$
$$\gamma_{qptm} \in \{0, 1\},\; \zeta_{ptm} \in \{0, 1\} \qquad \forall\, p, q \in \bar{P},\, t \in \bar{T},\, m \in \bar{M} \tag{4.36}$$

The setup variables γ_{qptm} receive an additional product parameter indicating *from* which product q a setup is performed *to* product p. Hence, when checking if a setup is performed to product p, the sum of the setup variables over all preceding products has to be calculated. With this extension, the objective function (4.27), the inventory flow equations (4.28), the capacity constraints (4.29) and the setup-to-production-linking conditions (4.30) are analogous to their CLSPL-BOPM counterparts (4.17) to (4.20). Constraints (4.31) are identical to (4.21). Equations (4.32) establish a kind of setup flow. If machine m is set up to product p in period t or if a setup carryover for p from the previous period is used, then the machine must be set up from p to (another) product r or the machine must carry over the setup state for p to the next period. This implies that every product—with the exception of the first and the last product per machine and period, respectively—has a predecessor and a successor. Conditions (4.33) generate the sequence of products per machine and period and thus eliminate sub-tours. The larger π_{ptm}, the later product p will be scheduled on machine m in period t. Whenever a setup from product q to product p is performed ($\gamma_{qptm} = 1$), the expression $P(1 - \gamma_{qptm})$ becomes zero, and thus $\pi_{ptm} \geq \pi_{qtm} + 1$ must hold true. This implies p is scheduled after q. Conditions (4.33) also prevent a setup to the same product ($\gamma_{pptm} \neq 1$ for all $p \in \bar{P}$, $t \in \bar{T}$, $m \in \bar{M}$). Conditions (4.34) and (4.35) are identical to (4.24) and (4.25), respectively. Conditions (4.36) include that the sequence variables π_{ptm} must be non-negative. They automatically become integer through conditions (4.33).

As mentioned above, one cannot expect to find an efficient algorithm that generates an optimal solution to these problems. For practical problem sizes, standard procedures like CPLEX already require a relatively long computation time to find a first (possibly very poor) solution. Thus, efficient heuristics are needed that do not guarantee optimality, but solve the problems sufficiently well in an acceptable amount of time. The following section presents such a solution approach.

4.4 Solution Approach

The core of the solution approach is a novel mixed integer programming model. In contrast to the CLSPL-BOPM or the CLSD-BOPM, the new model is no exact mathematical representation of the given lot-sizing problem. While some elements of the underlying problem and their interactions are modeled explicitly, others are approximated. For this reason, we call it a 'heuristic model'. The goal is to represent the real problem sufficiently well with a model that can be solved to optimality or near-optimality using standard mixed integer programming procedures even for practical problem sizes.

In order to find a solution to the original problem, the heuristic model is embedded in a lot-sizing and scheduling procedure. Within this procedure, the model is solved iteratively and its solutions are transformed to a solution of the original problem. This is done by generating values for the decision variables of the CLSPL-BOPM. Since the lot-sizing and scheduling procedure establishes a total order of all products per machine, it would also be possible to generate values for the decision variables of the CLSD-BOPM. However, the heuristic model itself does not generate a total sequence of the products and does not handle sequence-dependent setups. Thus, we present the algorithm as a solution procedure for the CLSPL-BOPM and not for the CLSD-BOPM.

We present the heuristic model and its behavior in Sect. 4.4.1. The lot-sizing and scheduling procedure is illustrated in Sect. 4.4.2. Section 4.4.3 concludes with some remarks for a rolling planning horizon.

4.4.1 A Novel Heuristic Model

4.4.1.1 Idea and Premises of the Model

The difficulty of the CLSPL-BOPM is caused by its high number of binary variables. The CLSP-BOPM has two binary variables for each (product $\times$ period $\times$ machine)-combination: One variable indicates that a setup is performed (γ_{ptm}) and another one that a setup is carried over from one period to the next one (ζ_{ptm}). The basic idea of the heuristic model is to avoid the binary variables. It uses integer variables instead. These variables indicate a number of machines, for example the number of machines that perform a setup for a certain product in a specific period. Since only one integer variable is needed to count all such machines, the number of non-continuous (i.e. 'difficult') variables is reduced substantially. This holds true especially when a large number of parallel machines is considered.

Only one aggregated machine is modeled explicitly in the capacity constraints. However, the parallel machines are accounted for implicitly by fixed-step setup costs and times: One setup has to be performed for every 'machine-quantity'—defined as a machine's capacity in product units for each product separately. If, for example, a machine can produce up to 100 units of product p per period and we produce 110 units, we have to set up two machines for product p. In order to do so, the model is based on two fundamental premises: The *first premise* is that we perform as few setups as possible to produce a certain production volume. This allows solutions to be found where, at most, one machine will be partially loaded by each product (premise 1′, Sect. 4.4.1.3): If a product uses more than one machine, all machines but one will be completely utilized by the product. The remaining production volume will be scheduled on a single additional machine. In general, this premise is in accordance with the behavior of a real shop floor: If a machine m is set up for any product p and not all units of p have been scheduled yet, we only set up a new machine for p after fully loading machine m.

The *second premise* is that the number of machines carrying over a setup for a certain product is limited by the product's production volume in machine-quantities: For each machine that may be fully loaded with the product, one setup can be carried over. This premise will be partially relaxed by the introduction of so-called 'dummy production volume', which makes it possible to find solutions that satisfy premise 2′ (Sect. 4.4.1.3): Only machines producing a single product in a period can carry-over their setup state to the next period. Machines that produce more than one product in a period are not able to carry over any setup state to the subsequent period. In many real-life problems, we will find high volume products utilizing complete machines and other low volume products sharing the remaining ones. Thus, in the heuristic model, the machines producing the high volume products will be able to carry over their setup state while the others will not. Since the latter produce a high number of products, it is relatively uncertain that the last product of a period and the first product of a subsequent period will be the same. We therefore assume that the misrepresentation of reality caused by this premise is tolerable.

Even with premises 1 and 2, capacity may be over- and underestimated, as is explained in Fig. 4.4. The example covers three machines and two periods. In period 1, machine 1 performs a setup and produces product a. Since it is the only product scheduled on machine 1, its setup state can be carried over to period 2 and production can continue without a new setup. Machine 2 is also set up for product a in period 1 and the remaining units are produced. Together, product a uses two machines in period 1. Machine 1 is completely utilized and machine 2 partially, in accordance with premise 1 because the production volume would not fit on a single machine. Machine 3 uses a setup carry-over from the previous period and produces product b. Product c remains to be produced. Since the total production volume of product c (including a setup) would fit on a completely free machine, the model reserves time for only one setup (premise 2). In fact, product c is loaded on machine 2 with a setup from product a. But because the machine is full before all units have been loaded, machine 3 is employed. In reality, we would have to perform a setup on machine 3 as well, but the model does not do so. Hence, capacity in period 1 is overestimated, because more units than possible are assigned to the machines.

On the other hand, the model underestimates capacity in period 2: Machine 3 continues production of product c. In reality, no setup for c has to be performed as c has been the last product in period 1. But, since the production volume of product c is less than a machine-quantity in period 1, the machine cannot carry over its setup state to period 2 (premise 2). The result is that the model schedules a setup for product c in period 2 and consumes capacity that is left unused in reality.

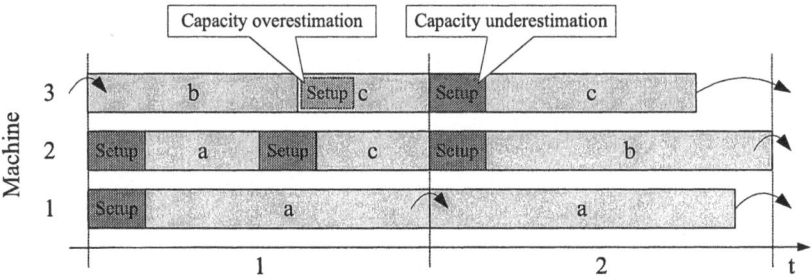

Fig. 4.4. Capacity over- and underestimation in the heuristic model

4.4.1.2 Model Formulation

We now present a MIP formulation of the heuristic model. In addition to the notation for the CLSPL-BOPM, we use the following symbols:

Parameters:

C^A Total per period capacity aggregated over all parallel machines, $C^A = M \cdot C$

C_p^c Capacity of a single machine in units of product p when the setup can be carried over from the previous period (machine-quantity with setup carry-over)

C_p^s Capacity of a single machine in units of product p when a setup has to be performed (machine-quantity with setup)

t_1 Index of first period, $t_0 = t_1 - 1$. E.g., $t_1 = 1$, $t_0 = 0$

T Index of last period, $\bar{T} = \{t_1 \ldots T\}$

ζ_p^0 Initial number of machines set up for product p

Variables:

x_{pt} Production volume of product p in period t

x_{pt}^D Dummy production volume of product p in period t

γ_{pt} Number of setups for product p in period t

$\hat{\gamma}_{pt}$ Number of machines used for product p in period t with setup carry-over from period $t - 1$

ζ_{pt} Number of machines with setup carry-over for product p between periods t and $t + 1$

ζ_{pt}^c Number of machines with setup carry-over for product p between periods t and $t + 1$ that have been carried over from period $t - 1$

ζ_{pt}^s Number of machines with setup carry-over for product p between periods t and $t + 1$ that have been set up in period t

Model HM (Heuristic Model)

$$\min \sum_{\substack{p \in \bar{P} \\ t \in \bar{T}}} \left(c_p^i y_{pt} + c_p^b b_{pt} + c_p^s \gamma_{pt} \right) \tag{4.37}$$

subject to

$$y_{p,\,t-1} - b_{p,\,t-1} + x_{pt} - d_{pt} = y_{pt} - b_{pt} \qquad \forall\, p \in \bar{P},\, t \in \bar{T} \tag{4.38}$$

$$\sum_{p \in \bar{P}} \left(t_p^u \left(x_{pt} + x_{pt}^D \right) + t_p^s \gamma_{pt} \right) \leq C^A \qquad \forall\, t \in \bar{T} \tag{4.39}$$

$$x_{pt} + x_{pt}^D \leq \hat{\gamma}_{pt} C_p^c + \gamma_{pt} C_p^s \qquad \forall\, p \in \bar{P},\, t \in \bar{T} \tag{4.40}$$

$$\hat{\gamma}_{pt} \leq \zeta_{p,\,t-1} \qquad \forall\, p \in \bar{P},\, t \in \bar{T} \tag{4.41}$$

$$\zeta_{pt} = \zeta_{pt}^c + \zeta_{pt}^s \qquad\qquad \forall\, p \in \bar{P},\, t \in \bar{T} \qquad (4.42)$$

$$x_{pt} + x_{pt}^D \geq \zeta_{pt}^c C_p^c + \zeta_{pt}^s C_p^s \qquad \forall\, p \in \bar{P},\, t \in \bar{T} \qquad (4.43)$$

$$\zeta_{pt} \geq \hat{\gamma}_{pt} + \gamma_{pt} - 1 \qquad\qquad \forall\, p \in \bar{P},\, t \in \bar{T} \qquad (4.44)$$

$$\sum_{p \in \bar{P}} \zeta_{pt} \leq M \qquad\qquad \forall\, t \in \bar{T} \qquad (4.45)$$

$$\hat{\gamma}_{pt} + \gamma_{pt} \leq M \qquad\qquad \forall\, p \in \bar{P},\, t \in \bar{T} \qquad (4.46)$$

$$\zeta_{pt}^c \leq \hat{\gamma}_{pt} \qquad\qquad \forall\, p \in \bar{P},\, t \in \bar{T} \qquad (4.47)$$

$$\zeta_{pt}^s \leq \gamma_{pt} \qquad\qquad \forall\, p \in \bar{P},\, t \in \bar{T} \qquad (4.48)$$

$$M - \sum_{p \in \bar{P}} \hat{\gamma}_{pt} \geq \sum_{p \in \bar{P}} \zeta_{pt}^s \qquad \forall\, t \in \bar{T} \qquad (4.49)$$

$$b_{pT} = 0 \qquad\qquad \forall\, p \in \bar{P} \qquad (4.50)$$

$$b_{pt_0} = b_p^0,\ y_{pt_0} = y_p^0,\ \zeta_{pt_0} = \zeta_p^0 \qquad \forall\, p \in \bar{P} \qquad (4.51)$$

$$\begin{aligned} b_{pt} \geq 0,\ x_{pt} \geq 0,\ y_{pt} \geq 0,\ \gamma_{pt} \in \mathbb{N}^0, \\ \hat{\gamma}_{pt} \in \mathbb{N}^0,\ \zeta_{pt} \in \mathbb{N}^0,\ \zeta_{pt}^c \in \mathbb{N}^0,\ \zeta_{pt}^s \in \mathbb{N}^0 \end{aligned} \qquad \forall\, p \in \bar{P},\, t \in \bar{T} \qquad (4.52)$$

The model uses two kinds of production variables. While variables x_{pt} indicate the production volume of product p in period t, the auxiliary variables x_{pt}^D represent the so-called 'dummy production volume'. Dummy production volume is not really produced and may therefore not be used to fulfil demand or put on inventory. It is only used to allow a setup carry-over for products that do not completely utilize a machine with real production volume, as formulated in conditions (4.43).

As in the CLSP-BOPM, the objective function (4.37) minimizes all related costs consisting of inventory holding costs, back-order costs, and setup costs. Equations (4.38) are inventory flow conditions. Capacity constraints are handled by conditions (4.39). The total processing and dummy processing time plus setup time may not exceed the aggregated capacity per period.

Conditions (4.40) ensure that it is only possible to produce a product with a certain volume if the according number of machines are set up for it. When setup times are considered, the machine capacity differs if a setup has to be performed in the period. The number of machines producing product p in period t with a setup carry-over from period $t-1$ is denoted by $\hat{\gamma}_{pt}$. Each of the $\hat{\gamma}_{pt}$ machines can produce C_p^c units, whereas each of the γ_{pt} machines has to perform a setup for product p and can produce C_p^s units in the remaining time. In realistic settings, $C_p^s \leq C_p^c$ will hold true for all p, but this is not necessary for the model.

A machine producing a product without a preceding setup in the same period must have its setup carried over from a previous period. On the other hand, a machine carrying over a setup for a product does not have to produce this product in the subsequent period. The number of machines carrying over a setup for product p from period $t-1$ to t is denoted by $\zeta_{p,t-1}$. Conditions (4.41) enforce $\zeta_{p,t-1}$ to be greater or equal to $\hat{\gamma}_{pt}$, meaning we may only use as many machines with a setup carry-over as there are machines carrying over a setup.

Equations (4.42) calculate the total number of machines carrying over a setup for a product. Two kinds of machines can carry over a setup for product p from period t to $t+1$: Machines whose setup states have already been carried over from period $t-1$ and machines that have been set up for p in period t. Their numbers are denoted by ζ_{pt}^c and ζ_{pt}^s, respectively. This differentiation is necessary because conditions (4.43) enforce that only machines being completely utilized producing a single product in a period—either with real or with dummy production—can carry over their setup state to the next period (premise 2). Since the capacity depends on the fact of whether a setup has to be performed (C_p^s units) or can be carried over from the previous period (C_p^c units), both cases have to be considered separately. Conditions (4.44) state that all machines but one producing a product p in period t must carry over their setup state to period $t+1$. Together with conditions (4.42) and (4.43), this implies that the production volume is produced with as few setups as possible (premise 1). Conditions (4.45) require each machine to carry over a setup state for only one product, and conditions (4.46) imply that no more than the total number of parallel machines—either with a setup or a setup carry-over—can produce a product during a period.

As seen in conditions (4.41), not all machines with a specific setup carry-over must produce the product in the subsequent period. On the other hand, a setup may only be carried over if a machine produces the product in the period—with regular or dummy production. Enforcing this, conditions (4.47) and (4.48) state that no more than the number of machines used for production can carry over a setup. Conditions (4.47) ascertain this for machines used without a setup—i.e. machines with a setup carry-over from a previous period—and conditions (4.48) for machines that have been set up in the actual period.

Conditions (4.49) are motivated by premise 2': Machines carrying over a setup for product p from period t to $t+1$ may not produce any other product in period t. Hence, if such a machine is being set up for p in (more precisely: at the beginning of) period t, it cannot use a possible setup carry-over for another product from $t-1$ to t for production in t. Conditions (4.49) enforce this: On the left hand side, we have the number of machines that do *not* use a setup carry-over in period t. Only those machines can both be set up in t and also carry over their setup to $t+1$, which is denoted by the sum on the right hand side.

Equations (4.50) state that, at the end of the planning horizon, all demands must have been met. This is achieved by fixing the last period's back-order variables to zero for all products. Equations (4.51) initialize the back-order and inventory volumes and set the initial setup states. Finally, conditions (4.52) enforce non-negative and integer variables, respectively.

For practical purposes, solving the model and handing the results—i.e. the production volumes and number of machines for each product—to the shop floor might be sufficient, especially if the parameters used in the lot-sizing model are stochastic in reality. For example, in practical settings, process times or machine downtimes are very often stochastic, but are assumed to be deterministic in most lot-sizing procedures. In this case, those effects may cover the approximate estimation of capacity in the heuristic model and the shop floor can immediately work with a solution generated by the heuristic model. If this is not the case and an exact solution for the original problem is needed, a solution of the heuristic model has to be transformed using the scheduling procedure described in Sect. 4.4.2.

4.4.1.3 Model Behavior

When a setup is to be scheduled in the original CLSPL-BOPM, it is necessary that the respective setup time is not larger than the capacity per machine. If this is not the case, the setup cannot be performed because the capacity conditions (4.19) would be violated. In the heuristic model, the capacity is only checked on an aggregate level. When a machine is overloaded by a setup that is longer than the machine's capacity, another machine could remain idle and compensate the overloading. Thus, the respective capacity conditions (4.39) would still be fulfilled. Nevertheless, if the setup time for a product p is longer than the capacity per machine, the machine-quantity C_p^s will be zero. Hence, in the specific period, no units can be loaded on a machine performing such a setup. On the other hand, in subsequent periods, a setup carry-over can be used to produce product p. To avoid such behavior, three additional constraints could be added:

$$\gamma_{pt} \leq F \cdot \left(1 - \gamma_p^i\right) \qquad \forall\, p \in \bar{P},\ t \in \bar{T} \qquad (4.53)$$

$$t_p^s - C \leq F \cdot \gamma_p^i \qquad \forall\, p \in \bar{P} \qquad (4.54)$$

$$\gamma_p^i \in \{0, 1\} \qquad \forall\, p \in \bar{P} \qquad (4.55)$$

with the additional symbols

F Big number, $F \geq \max\left\{\max_{p \in \bar{P}}\left\{t_p^s\right\}, M\right\}$

γ_p^i Binary variable ($\gamma_p^i = 1$, if a setup for product p is impossible due to its setup time)

Notice that the variables γ_p^i are not complicating the model as they are fully determined by the instance data: Given any instance, the model has no choice on how to set them.

The constraints state that it is only possible to set up a machine for a product p if the setup time t_p^s is less than the capacity per machine C. However, we do not include those additional constraints and simply assume that each setup time is not longer than the capacity per machine. This is a standard assumption for big bucket models. In any case, if a setup time were longer than the capacity per machine, no feasible solution that schedules such a setup would exist in the sense of the CLSPL-BOPM.

In the remainder of this section, we explain how every solution of the model may be transformed into one that satisfies premises 1' and 2', that is:

premise 1': at most, one machine will be partially loaded by each product (an extension of premise 1)

premise 2': only machines producing a single product in a period can carry over their setup state to the next period (an extension of premise 2)

Consider the following case: A solution of model HM may indicate that a setup state is carried over from a period t to $t+1$ on a machine that performs a setup in t, but premise 2' dictates that another machine with a setup carry-over from $t-1$ shall carry over the setup to $t+1$ (ζ_{pt}^s is positive instead of ζ_{pt}^c). In model HM, a setup can be carried over as long as the production volume (including dummy production) is high enough to utilize the complete machine as given by parameters C_p^s and C_p^c (see conditions (4.43)). Hence, whenever a machine used in t with a setup carry-over from $t-1$ preserves the setup until $t+1$, also a machine performing a setup in t is able to carry over its setup state, as the latter has less capacity for production than the former ($C_p^s \leq C_p^c$). On the other hand, premise 2' might prevent machine 1 from carrying over its setup to $t+1$, if the machine produces another product in t before changing to p. Figure 4.5 gives an example for this situation: Machine 1 carries over a setup for product a even though it produces more than one product in period t ($\zeta_{at}^c = 0$ and $\zeta_{at}^s = 1$). Also premise 1' is violated as both machines 1 and 2 are only partially loaded by product a.

The solution can easily be transformed by shifting production volume from the machine performing a setup in t to the machine using a setup carry-over from $t-1$ (in the example, that is from machine 1 to 2). The latter machine will be completely utilized by the product and can carry over its setup state to the subsequent period in accordance with premises 1' and 2'. Further products being originally assigned to this machine can be loaded on the other machine. Figure 4.6 shows the result for the example.

An analogous example is depicted in Fig. 4.7. This time, the production volume for product a on machine 1 is not large enough to fill up machine 2

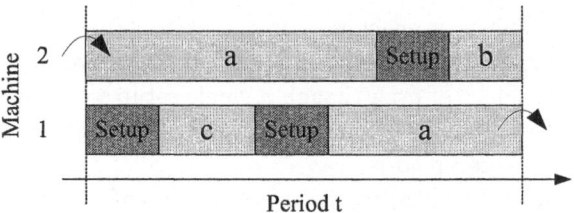

Fig. 4.5. Solution not satisfying premises 1′ and 2′

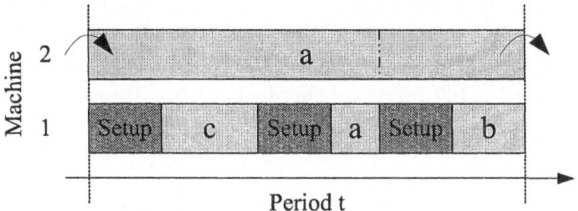

Fig. 4.6. Transformed solution

after the transformation as above, and the setup on machine 2 is carried over to period $t + 1$, even though the machine is not completely utilized. Thus, premises 1′ and 2′ are satisfied, but premise 2 is not. However, this noncompliance with premise 2 is tolerable, as the remaining capacity cannot be used for anything else—in fact, the model will have used the capacity for

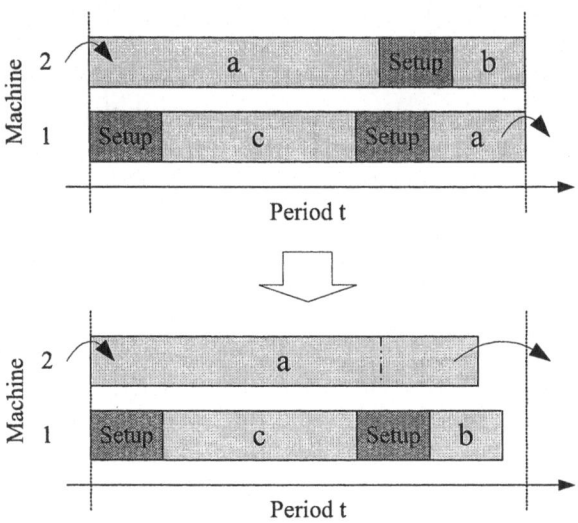

Fig. 4.7. Initial and transformed solution with leftover capacity

a setup (for product a on machine 1 in the example) that is not performed in the transformed solution. Notice as well that the initial solution scheduled two setups for product a, even though a single setup would be sufficient. As such, the initial solution also violates premise 1 and cannot be optimal in the presence of positive setup costs.

Adding constraints (4.56) to the model would enforce premises 1′ and 2′ directly:

$$\zeta_{pt}^c \geq \min\{\hat{\gamma}_{pt},\ \zeta_{pt}\} \qquad\qquad \forall\, p \in \bar{P},\ t \in \bar{T} \qquad (4.56)$$

Constraints (4.56) deal with the case that a setup carry-over for a product p from period t to $t+1$ is planned and the decision has to be made which machine to use for it. The constraints state that when there is a choice between a machine that performs a setup for p in t and a machine that uses a setup carry-over from $t-1$, the latter machine will always carry over the setup to period $t+1$: The number of machines using a setup carry-over from $t-1$ and preserving it to $t+1$ (ζ_{pt}^c) will always be greater than the minimum of the number of machines using a setup carry-over ($\hat{\gamma}_{pt}$) and the number of machines carrying over a setup to the next period (ζ_{pt}). In other words, each setup carry-over shall be performed by a machine using a setup carry-over from the previous period as long as there are such machines. Since only machines being fully utilized by a single product can carry over their setup state to the next period, this also implies that machines using a setup carry-over from the previous period will be fully utilized before another machine is set up for the product, enforcing both premises 1′ and 2′.

Nevertheless, the constraints are not added to the model formulation because they are non-linear and would therefore substantially increase the solution time of the model. As described above, any solution violating these constraints may be transformed into one where they are respected. The scheduling procedure will do this automatically. Thus, the inclusion of constraints (4.56) is not necessary.

The scheduling procedure must also take into consideration that machines that produce a single product for the complete period do not have to preserve their setup state to the next period: If the setup state is not used in the subsequent period, the model does not need to set the setup carry-over variable ζ_{pt} accordingly, because there is no constraint forcing it to do so. That is, $\zeta_{pt} = 0$ may hold true even though $x_{pt_1} + x_{pt_1}^D \geq C_p^c$. In any case, conditions (4.44) limit this to, at most, one machine per product and period.

4.4.1.4 Valid Inequalities

Valid inequalities are relationships within a model that hold true without being explicitly enforced by constraints. However, they are sometimes added to model formulations in order to speed up solution times. They explicitly show a solver that certain regions of the solution space are infeasible and thus do not have to be investigated. A number of valid inequalities hold true for the heuristic model with respect to premises 1' and 2':

For all products together, the number of scheduled setups per period is limited by $M + P - 1$. This is because for each product, at most, one machine may be partially loaded (premise 1'), leading to a maximum of P setups for these volumes. Let's assume these volumes are loaded on m machines. If there are no such machines (i.e. $m = 0$), the number of setups is limited by M ($\leq M - 1 + P$ since $P \geq 1$) because each of the M machines can perform, at most, one setup. If $m \geq 1$, $n := M - m$ machines are left. Each of them is loaded by, at most, one product. Since n is limited by $M - 1$, this leads to, at most, $M - 1$ setups for machines producing only a single product, resulting in a maximum number of setups of $M + P - 1$.

Thus, the following inequality holds true:

$$\sum_{p \in \bar{P}} \gamma_{pt} \leq M + P - 1 \qquad\qquad \forall\, t \in \bar{T} \qquad\qquad (4.57)$$

The left hand side counts the number of setups for all products in period t. It is possible to tighten this constraint. In general, the number of setups per period will be less than $M + P - 1$ because machines may be using a setup carry-over from the previous period. For each setup carry-over used, one less setup will be made. This leads to the following inequality:

$$\sum_{p \in \bar{P}} (\gamma_{pt} + \hat{\gamma}_{pt}) \leq M + P - 1 \qquad\qquad \forall\, t \in \bar{T} \qquad\qquad (4.58)$$

Clearly, (4.58) is tighter than (4.57). The left hand side counts the number of setups plus the number of machines using a setup carry-over from the previous period. Together, this is the number of production lots per period. Hence, (4.58) implies that the number of production lots per period as scheduled by the heuristic model is limited by $M + P - 1$.

If all machines carry over their setup state to the next period, no machine may produce more than one product (premise 2'). That means that the number of production lots per period must be less than or equal to the number of machines, leading to a variation of inequality (4.58):

$$\sum_{p\in\bar{P}} (\gamma_{pt} + \hat{\gamma}_{pt}) \leq M + P\left(M - \sum_{p\in\bar{P}} \zeta_{pt}\right) \qquad \forall\, t \in \bar{T} \qquad (4.59)$$

If all machines carry over their setup state to the next period, $\sum_{p\in\bar{P}} \zeta_{pt} = M$ holds true and the last term on the right hand side of (4.59) becomes zero. Thus, the number of production lots (as calculated on the left hand side) must be less than or equal to M. In fact, even equality will hold true because only machines producing (dummy or non-dummy) volumes can carry over their setup. For $\sum_{p\in\bar{P}} \zeta_{pt} < M$, (4.59) is also valid as it is less tight than (4.58).

If all machines carry over their setup state to the next period, each of them produces a single product only (premise 2'). If all these machines use a setup carry-over from the previous period, no setup at all can be scheduled in the actual period. This results in the following inequality:

$$\sum_{p\in\bar{P}} \gamma_{pt} \leq P\left(M - \sum_{p\in\bar{P}} \zeta_{pt}^c\right) \qquad \forall\, t \in \bar{T} \qquad (4.60)$$

The right hand side of (4.60) will become zero if all machines use a setup carry-over from the previous period, produce only a single product, and carry over their setup state to the next period. The left hand side counts the number of setups, which must be zero in this case. The inequality also holds true for $\sum_{p\in\bar{P}} \zeta_{pt}^c < M$, as will be proven below:

For all $a,\, b \in \mathbb{N}^{>0}$: $b\cdot a \geq a - 1 + b$ holds true.

Proof: If $a = b = 1$, $1 \geq 1$ holds true. If $a > 1$ and $b = 1$, $b\cdot a = a \geq a - 1 + 1 = a$ holds true. The case $b > 1$ and $a = 1$ is analogous. If $a > 1$ and $b > 1$, $b\cdot a \geq a - 1 + b \iff b\cdot a - a \geq b - 1 \iff (b-1)a \geq b - 1 \overset{b\geq 1}{\iff} a \geq 1$. This is true since $a > 1$, qed.

With $b = P$ and $a = M - \sum_{p\in\bar{P}} \zeta_{pt}^c$, we get

$$P\left(M - \sum_{p\in\bar{P}} \zeta_{pt}^c\right) \geq M - \sum_{p\in\bar{P}} \zeta_{pt}^c + P \qquad \forall\, t \in \bar{T}. \qquad (4.61)$$

The left hand side of (4.61) is the right hand side of (4.60). On the right hand side of (4.61), $M - \sum_{p\in\bar{P}} \zeta_{pt}^c$ is the maximum number of machines that can perform setups (because $\sum_{p\in\bar{P}} \zeta_{pt}^c$ machines use a setup carry-over and produce a single product only). Thus, with the explanations given for inequality (4.57), the right hand side of (4.61) represents a limit on the number of setups. Together, it follows that inequality (4.60) is valid also for $\sum_{p\in\bar{P}} \zeta_{pt}^c < M$.

We have not added these inequalities to the model formulation as preliminary tests have shown that they appear to increase the solution time.

However, they may be used as a basis to create other valid inequalities that may shorten the solution time of the model.

4.4.2 Solution Procedure

We will now present a solution procedure that comprises the heuristic model as its core. The result will be a solution for the original problem represented by the CLSPL-BOPM, given by values for its decision variables. It is sufficient to determine values for x_{ptm}, γ_{ptm} and ζ_{ptm}, as the other variables may be calculated using the inventory flow equations (4.18). Notice that variables of the CLSP-BOPM can be distinguished from the ones of the heuristic model by the additional index m indicating the machine.

When the heuristic model does not find any solution to a given instance due to capacity restrictions, we would like to know how much additional capacity would be needed. This information is used in Sect. 4.4.2.3. In order to get this information, we augment the model by an overtime variable C^O for the first period t_1 only. Also, for purposes explained in Sect. 4.4.2.3, we add a period index t to the capacity parameter C^A. The new capacity constraints replacing (4.39) are shown in (4.62) and (4.63).

$$\sum_{p\in\bar{P}} \left(t_p^u \left(x_{pt_1} + x_{pt_1}^D\right) + t_p^s \gamma_{pt_1}\right) \leq C_{t_1}^A + C^O \tag{4.62}$$

$$\sum_{p\in\bar{P}} \left(t_p^u \left(x_{pt} + x_{pt}^D\right) + t_p^s \gamma_{pt}\right) \leq C_t^A \qquad \forall\, t \in \bar{T} \setminus t_1 \tag{4.63}$$

Since the original problem does not include the option to use overtime, we treat a solution that uses overtime as an infeasible solution. Therefore, we would like to use overtime only if absolutely necessary and adjust the objective function: Overtime is punished by prohibitive costs c^o, which are set to a high value. The new objective function replacing (4.37) is shown in (4.64).

$$\min \sum_{\substack{p\in\bar{P}\\t\in\bar{T}}} \left(c_p^i y_{pt} + c_p^b b_{pt} + c_p^s \gamma_{pt}\right) + c^o C^O \tag{4.64}$$

Overtime costs may be set to

$$c^o = 10\,000 \cdot \sum_{\substack{p\in\bar{P}\\t\in\bar{T}}} d_{pt} \cdot \sum_{p\in\bar{P}} \left(c_p^i + c_p^b\right) + \sum_{p\in\bar{P}} c_p^s \cdot M \cdot T\,. \tag{4.65}$$

Finally, we have to ensure non-negativity for the overtime variables:

$$C^O \geq 0 \tag{4.66}$$

We call the new **model HMO** (Heuristic Model with Overtime). It consists of objective function (4.64) and equations (4.38), (4.40)–(4.52), (4.62)–(4.63) and (4.66).

The solution procedure comprises a lot-sizing and a scheduling component. The lot-sizing procedure solves model HMO. Afterwards, the scheduling procedure schedules the product units on the parallel machines.

4.4.2.1 Lot-Sizing Procedure

The lot-sizing procedure iterates in a forward period-by-period manner. For each period t_1 from the chronologically ordered set $(1, 2, \ldots, T)$, an instance of model HMO covering periods $t_1, t_1 + 1, \ldots, T$ is solved to optimality or near-optimality using standard procedures (CPLEX). If a solution without overtime is found, only the results for period t_1 are analyzed and production quantities and machines assignments for this period are fixed. Afterwards, t_1 is incremented to the subsequent period and the procedure continues with the next iteration. If a solution uses overtime and is hence infeasible in the sense of the original problem, we perform a period backtracking as described in Sect. 4.4.2.3. Figure 4.8 outlines the lot-sizing procedure.

Figure 4.9 illustrates an ideal run of the lot-sizing procedure—i.e. no backtracking step has to be performed. In iteration 1, the lot-sizing problem covers all periods from 1 to $T = 5$. With each iteration, the lot-sizing instances become smaller, as t_1 is increased and the former first period is excluded. After solving an instance of model HMO, the scheduling procedure is invoked for period t_1.

In Sect. 4.2, we outlined a taxonomy of lot-sizing procedures presented by Maes and Van Wassenhove (1988). Our procedure combines various aspects of different classes in this taxonomy. Firstly, it is a period-by-period heuristic. However, the production volumes to be produced in each period are determined in a more sophisticated way than by a simple priority rule. We use a mathematical programming procedure instead and solve model HMO. In this

---

For each period $t_1 = 1, 2, \ldots, T$

1. Solve HMO covering periods $t_1, t_1 + 1, \ldots, T$
2. If HMO could be solved without overtime
    - Calculate production quantities and machine assignments for period t_1 (scheduling procedure)
   else
    - Period backtracking to reach feasibility

Fig. 4.8. Lot-sizing procedure

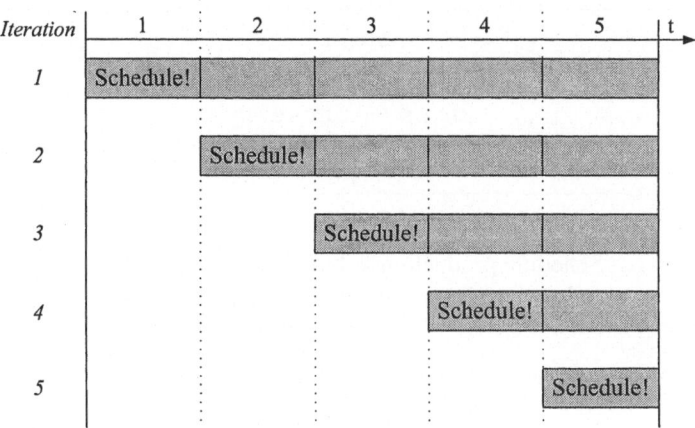

Fig. 4.9. A run of the lot-sizing procedure without period backtracking

sense, we couple common-sense with mathematical programming. Secondly, during the period-by-period approach, the feasibility check is accomplished using both an a-priori anticipation of future bottlenecks *and* a feedback procedure: Solving model HMO anticipates future capacity demands implicitly, and the feedback procedure is represented by the period backtracking. With the model HMO as the core of the procedure, it should be easy to extend and adapt the algorithm to different problem variants.

4.4.2.2 Scheduling Procedure

Taking the (near-)optimal solution of the actual instance of model HMO, the scheduling procedure reads the values of the variables x_{pt_1}, $x_{pt_1}^D$, $\hat{\gamma}_{pt_1}$ and ζ_{pt_1} for the current period t_1 and executes three steps which are summarized in Fig. 4.10. Step 1 and 2 schedule production volumes that can be assigned to machines using the results of model HMO directly. In step 3, an additional scheduling model is applied to schedule the remaining production volumes.

Step 1 schedules production volumes that use a setup carry-over. In the example illustrated in Fig. 4.4, this is product b on machine 3 in period 1 and product a on machine 1 in period 2.

In general, for each product p, we reserve a machine for all production volumes using a setup carry-over from period $t_1 - 1$. The number of such machines is given by $\hat{\gamma}_{pt_1}$. Let m be a machine that carried over a setup for product p from period $t_1 - 1$ to t_1. Since no setup has to be performed, the machine capacity is C_p^c units. If the remaining production volume $x_{pt_1} + x_{pt_1}^D$ exceeds this capacity or if $\zeta_{pt_1} > 0$ (see Sect. 4.4.1.3), we will only load product p on machine m in t_1. As dummy volumes do not exist in reality, we set

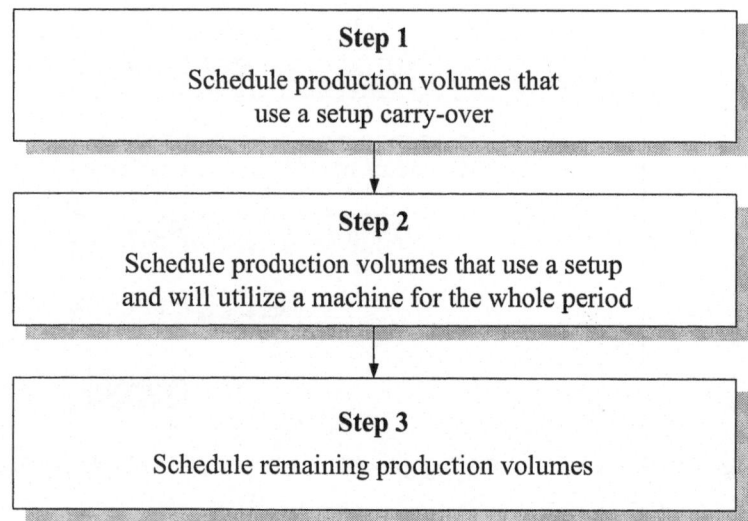

Fig. 4.10. Scheduling procedure overview

x_{pt_1m} of the CLSPL-BOPM to the minimum of x_{pt_1} and C_p^c units. Afterwards, we reduce the remaining production and dummy production volume by the amount just scheduled. The machine is declared full and will carry over its setup for p to period $t_1 + 1$. Therefore, the setup carry-over counter ζ_{pt_1} is reduced by one (if it was positive before, see Sect. 4.4.1.3). If, on the other hand, the remaining production volume $x_{pt_1} + x_{pt_1}^D$ does not allow the full loading of a machine and $\zeta_{pt_1} = 0$, only the remaining no-dummy volume x_{pt_1} is scheduled and the machine may still be used in step 3 of the scheduling procedure. Figure 4.11 illustrates the details of this step.

In step 2, we schedule product volumes that use a setup and will utilize a machine for the whole period.

In the example of Fig. 4.4, product a on machine 1 in period 1 and product b on machine 2 in period 2 are scheduled this way. Also, product c on machine 3 in period 2 is scheduled in step 2 as the arrow at the right of machine 3 indicates that the setup is carried over to the next period. This means that the remaining capacity of machine 3 is filled with dummy production volume (see constraints (4.43)).

Machines being utilized by a single product for the whole period will have their setup state preserved for period $t_1 + 1$. So, for each product p and its remaining setup carry-overs to period $t_1 + 1$ as indicated by ζ_{pt_1}, we reserve a yet unused machine m and assign the appropriate production volume to fully load the machine. Since a setup has to be scheduled, the machine capacity is given by C_p^s and the CLSP-BOPM variables γ_{pt_1m} are set to one, indicating

For each product $p \in \bar{P}$

While there are machines using a setup carry-over from $t_1 - 1$ for p in t_1 ($\hat{\gamma}_{pt_1} \geq 1$)
1. Select a machine m that carried over a setup for p from $t_1 - 1$ to t_1
2. Declare m used
3. If $x_{pt_1} + x_{pt_1}^D \geq C_p^c$ or $\zeta_{pt_1} > 0$
   - Load machine m with product p: Set $x_{pt_1m} = \min\left\{C_p^c, x_{pt_1}\right\}$
   - Decrease remaining production volume:
     Set $x_{pt_1} = x_{pt_1} - x_{pt_1m}$ and
     $x_{pt_1}^D = \max\left\{x_{pt_1}^D - \max\left\{C_p^c - x_{pt_1}, 0\right\}, 0\right\}$
   - Declare m full
   - Reduce remaining machines with setup carry-over to period $t_1 + 1$: Set $\zeta_{pt_1} = \max\left\{\zeta_{pt_1} - 1, 0\right\}$

   else
   - Load machine m with product p: Set $x_{pt_1m} = x_{pt_1}$
   - Decrease remaining production volume: Set $x_{pt_1} = 0$
4. Reduce remaining machines using setup carry-over for p in t_1:
   Set $\hat{\gamma}_{pt_1} = \hat{\gamma}_{pt_1} - 1$

Fig. 4.11. Scheduling procedure (Step 1): Machines using setup carry-over

For each product $p \in \bar{P}$

While there are machines with a setup carry-over to period $t_1 + 1$ ($\zeta_{pt_1} \geq 1$)
1. Select a yet unused machine m
2. Declare m used and full
3. Schedule a setup for product p on machine m: Set $\gamma_{pt_1m} = 1$
4. Load machine m with product p: Set $x_{pt_1m} = \min\left\{C_p^s, x_{pt_1}\right\}$
5. Decrease remaining production volume:
   Set $x_{pt_1} = x_{pt_1} - x_{pt_1m}$ and
   $x_{pt_1}^D = \max\left\{x_{pt_1}^D - \max\left\{C_p^s - x_{pt_1}, 0\right\}, 0\right\}$
6. Reduce remaining machines with setup carry-over to period $t_1 + 1$: Set $\zeta_{pt_1} = \zeta_{pt_1} - 1$

Fig. 4.12. Scheduling procedure (Step 2): Full machines with setup

a setup. Variables x_{pt_1m} are set in an analogous manner to step 1 and the production and dummy production volumes are updated accordingly. Notice that, with the exception of the case mentioned in Sect. 4.4.1.3, as long as $\zeta_{pt_1} \geq 1$, $x_{pt_1} + x_{pt_1}^D \geq C_p^s$ will hold true. Figure 4.12 shows the step in detail.

Step 3 schedules the remaining production volumes. These are production volumes that neither use a setup carry-over nor utilize a machine for the whole period. In the example shown in Fig. 4.4, this holds true for product a on machine 2 and product c on machines 2 and 3, all in period 1. Solutions of model HMO do not indicate on which machines and in which sequence to schedule those quantities—in fact, no feasible solution might exist as in the case of period 1 of the example in Fig. 4.4.

In general, a heuristic procedure could be invoked to find a machine and time assignment for the remaining volumes. However, the problem instances are relatively small. They only contain machines that are not yet fully utilized and products with a remaining production volume. Hence, we apply an additional mixed integer programming scheduling model that may be solved using standard procedures (CPLEX). Even if just near-optimal solutions are obtained, the overall objective value for the original lot-sizing and scheduling problem will only be affected marginally. In addition to the notation for the CLSPL-BOPM, we use the following symbols:

Parameters:

C_m	Remaining capacity of machine m after step 2 of the scheduling procedure
f	Setup costs devaluation factor, e.g. $f = 0.00001 \cdot \frac{\min_{p \in P}\{c_p^b\}}{\max_{p \in P}\{c_p^s\}}$
M'	Number of remaining (i.e. not yet fully utilized) parallel machines, $\bar{M}' = \{1 \dots M'\}$
q_p	Remaining production volume of product p after step 2 of the scheduling procedure
s_p	Setup carry-over bonus for product p ($s_p = c_p^s$, if $t_1 < T$ and the optimal solution of model HMO scheduled a setup for p in $t_1 + 1$)
ζ_{pm}^0	Initial setup state of product p on machine m after step 2 of the scheduling procedure ($\zeta_{pm}^0 = 1$, if machine m is initially set up for product p)

Variables:

π_{pm}	Sequencing variable indicating the ordinal position of product p on machine m. The larger π_{pm}, the later product p is scheduled on m.
x_{pm}	Fraction of the remaining production volume q_p of product p produced on machine m
γ_{pqm}	Binary setup variable indicating a setup from product p to product q on machine m ($\gamma_{pqm} = 1$, if a setup is performed from product p to product q on machine m)
ζ_{pm}	Binary setup carry-over variable for product p on machine m ($\zeta_{pm} = 1$, if the setup state for product p on machine m is carried over from period t_1 to $t_1 + 1$)

Model Scheduler

$$\min \sum_{p \in \bar{P}} c_p^b q_p \cdot \left(1 - \sum_{m \in \bar{M}'} x_{pm}\right) + f \sum_{\substack{p,\,q \in \bar{P} \\ m \in \bar{M}'}} c_p^s \gamma_{pqm} - f \sum_{\substack{p \in \bar{P} \\ m \in \bar{M}'}} s_p \zeta_{pm} \qquad (4.67)$$

subject to

$$\sum_{m \in \bar{M}'} x_{pm} \leq 1 \qquad\qquad \forall\, p \in \bar{P} \qquad\qquad (4.68)$$

$$\sum_{p,\,q \in \bar{P}} t_p^s \gamma_{pqm} + \sum_{p \in \bar{P}} t_p^u q_p x_{pm} \leq C_m \qquad \forall\, m \in \bar{M}' \qquad (4.69)$$

$$x_{pm} \leq \zeta_{pm}^0 + \sum_{q \in \bar{P}} \gamma_{qpm} \qquad \forall\, p \in \bar{P},\ m \in \bar{M}' \qquad (4.70)$$

$$\sum_{p \in \bar{P}} \zeta_{pm} \leq 1 \qquad\qquad \forall\, m \in \bar{M}' \qquad (4.71)$$

$$\zeta_{pm}^0 + \sum_{\substack{q \in \bar{P} \\ q \neq p}} \gamma_{qpm} = \zeta_{pm} + \sum_{\substack{q \in \bar{P} \\ q \neq p}} \gamma_{pqm} \qquad \forall\, p \in \bar{P},\ m \in \bar{M}' \qquad (4.72)$$

$$\pi_{qm} \geq \pi_{pm} + 1 - P\,(1 - \gamma_{pqm}) \qquad \forall\, p, q \in \bar{P},\ m \in \bar{M}' \qquad (4.73)$$

$$\pi_{pm} \geq 0,\ x_{pm} \geq 0,\ \gamma_{pqm} \in \{0, 1\},\ \zeta_{pm} \geq 0 \quad \forall\, p, q \in \bar{P},\ m \in \bar{M}' \qquad (4.74)$$

The objective is to produce the complete volume given by the solution of model HMO. It might not be possible to do so because model HMO only estimates the capacity consumption. Hence, the first term of the objective function (4.67) minimizes the additional back-order costs incurred by volume that cannot be scheduled because of the capacity misestimation. Among the solutions with the lowest back-order costs, we would like to minimize setup costs. Therefore, the second term punishes setup costs, devaluated by factor f. The devaluation factor f must be sufficiently small to give absolute priority to the first objective, i.e. to generate as little additional back-order as possible. The third term takes into account that a solution of model HMO might schedule setups in the subsequent period $t_1 + 1$. If possible, it would be advantageous to schedule those products last in the actual period t_1 and carry over the setup to period $t_1 + 1$. For that reason, all products having a scheduled setup in period $t_1 + 1$ receive a setup carry-over bonus of their respective setup costs in the scheduling model, again devaluated by factor f. In this way, we encourage solutions leading to low setup costs in the actual and the next period.

An alternative scheduling procedure would be not to use a devaluation factor—i.e. to set $f = 1$. The objective function would then weight back-order and setup costs equally and minimize them (see Sect. 4.4.2.5).

Conditions (4.68) ensure that production does not exceed the given volume by restricting variables x_{pm} to [0, 1]. Conditions (4.69) limit the time for processing and setups to the remaining machine capacity C_m left after step 2 of the scheduling procedure. Conditions (4.70) ascertain that production of a product can only take place if the specific machine is set up for it.

Constraints (4.71)–(4.73) are based on constraints (4.31)–(4.33) from the CLSD: Conditions (4.71) state that each machine can carry over, at most, one setup state to period $t_1 + 1$. Equations (4.72) contribute the setup flow: If a machine is initially set up for a product p or a setup to product p is performed, the machine must either carry over this setup to the next period or a setup from p to another product q must be performed. Conditions (4.73) establish a total sequence of all products per machine and avoid sub-tours. If a setup from product p to product q is scheduled on machine m, the ordinal position of q on m must be higher than the one of p, meaning that p is produced before q. Finally, conditions (4.74) declare the setup variables γ_{pqm} binary and all other variables positive. Due to the model formulation, variables π_{pm} and ζ_{pm} automatically become binary.

The presented formulation contains all original products. In order to keep the model small, a better approach would be to include only products with a remaining production volume. There are two obstacles to this: Firstly, a machine might be initially set up for a product that has no remaining production volume—in fact, all machines will have such an initial setup state—and secondly, also those products that do not have a remaining production volume might be associated with a setup carry-over bonus. To remedy the situation, two sets of products should be maintained: One with all products with remaining demand, and one with all original products. Then, as an example, the first index p of the setup variables γ_{pqm} would be running over all original products and the second index q only over those products with a remaining production volume. To keep the presentation simple, we do not show this straightforward but slightly cumbersome extension. Further, notice that, if $t_1 < T$, the scheduling procedure needs variables $\gamma_{p,\,t_1+1}$ of the (near-)optimal HMO solution in addition to the variables mentioned at the beginning of this section in order to determine the products that will receive a setup carry-over bonus.

After calculating the needed parameters, step 3 of the scheduling procedure solves the scheduling model to optimality or near-optimality using standard procedures (CPLEX). The results are read and the appropriate production and setup variables x_{pt_1m} and γ_{pt_1m} of the CLSP-BOPM are set.

Besides determining the product/machine assignments, the scheduling model also sequences the products on each machine, delivering the setup carry-over parameters needed for the next iteration of the lot-sizing procedure. The scheduling procedure combines this information with the results from step 1 and 2 and sets the setup carry-over variables ζ_{pt_1m} of the CLSP-BOPM. Further, it counts the produced units per product and sets the back-order and inventory variables (b_{pt_1} and y_{pt_1}). While inventory is always due to the fact that products are produced before their period of demand in the solution of model HMO ($y_{pt_1} > 0$ in the solution of model HMO), back-orders can originate from two causes: They may as well be given by the solution of model HMO directly ($b_{pt_1} > 0$ in the solution of model HMO), or the scheduling procedure was not able to produce the total production volume given by the HMO solution due to its wrong capacity estimation ($b_{pt_1} = 0$ in the solution of model HMO, but $b_{pt_1} > 0$ after the scheduling procedure).

The new initial back-order and inventory volumes for the next lot-sizing iteration are updated by setting $b_p^0 = b_{pt_1}$ and $y_p^0 = y_{pt_1}$.

This concludes one iteration of the lot-sizing and the scheduling procedure. The production volumes and time/machine assignments for the actual period t_1 are set. If $t_1 < T$, the lot-sizing procedure continues with the next period $t_1 + 1$. If $t_1 = T$ and there are no back-orders, a solution to the original problem has been found. Otherwise, a period backtracking is invoked.

4.4.2.3 Period Backtracking to Reach Feasibility

A period backtracking is executed whenever no feasible solution in the sense of the original problem can be found (and $t_1 > 1$). This occurs in two cases: In the first case, the lot-sizing procedure cannot find a solution to model HMO that does not use overtime. In the second case—which is only possible at the end of the planning horizon for $t_1 = T$—the scheduling procedure is unable to find a solution without back-ordering in period T. Both cases may only happen when the lot-sizing procedure has overestimated the capacity. In the first case, this happened in the last iteration for period $t_1 - 1$. The scheduling procedure could not schedule the given production volume and had to back-order (additional) units to period t_1. In period t_1, the lot-sizing procedure has to schedule more units than projected and cannot do so without using overtime. In the second case, the capacity misestimation happened in period $t_1 = T$ itself. While a period backtracking is always preceded by a capacity overestimation, a capacity overestimation need not necessarily entail a period backtracking.

The idea behind the period backtracking is as follows: When no feasible solution could be found for period t_1, the aggregate capacity of that period is reduced and the planning is redone starting one period earlier. The lot-sizing procedure will have the same number of machines but less aggregate capacity available—and find a solution with less capacity consumption in t_1, possibly by shifting production volumes to period $t_1 - 1$.

To prepare the period backtracking, we calculate C^f, the amount of additional capacity needed to find a feasible solution without overtime. In the first of the two cases mentioned above, this information can be read directly from the overtime variable C^O. In the second case, some calculations have to be performed: For each product p, we multiply the back-order volume b_{pT} with the specific process time t_p^u. Summed up for all products, we get the total additional capacity required for processing (C_p^f). We include setup times using the following approximation: For each machine m producing a product p with a positive back-order volume, we assume that no other product is produced on m. Hence, the machine may be utilized completely by product p and additional units of p might be loaded on m. The back-order volume of p is reduced by that amount summed up for all such machines. The remaining volume would have to be scheduled on further machines performing a setup. Their number is calculated by dividing the remaining back-order volume by the machine capacity in product units (C_p^s) and rounding up the result. Multiplied with the specific setup time and summed up over all products, we get an approximation of the total additional capacity needed for setups (C_s^f). C^f is then set to $C_p^f + C_s^f$.

The above method approximates the additionally needed setups by a lower bound: In general, the number of additional units on machines that already produce product p will be less, because machines might have been loaded with other products as well. The reason for this modus operandi is as follows: Less capacity will result in a lower production volume after period backtracking— and every product is eligible for a production volume reduction, including the products that are not 'responsible' for the capacity overestimation (i.e. the products whose production volumes as given by the solution of model HMO could be completely scheduled by the scheduling procedure). If such a product is produced on a machine that also produces product p (with $b_{pT} > 0$), more units of p might be scheduled in the new solution. By using the lower bound approximation, capacity is decreased in small steps in order not to exclude such potentially feasible solutions.

When re-starting the lot-sizing procedure from period $t_1 - 1$ with a reduced aggregate capacity in t_1, we would like to enforce a lower capacity usage in t_1. However, the former solution of model HMO for period $t_1 - 1$ might not have used the full aggregate capacity in t_1. Therefore, we must take the amount of capacity used by this solution as a basis for the reduction. Otherwise, the reduction might have no effect as it does not impose a tight constraint on the problem. So, whenever a solution of model HMO has been found in a period t, we calculate its capacity usage in period $t+1$ (C_{t+1}^{used}). This information is used when a period backtracking is performed: We reduce the aggregate capacity of period t_1 by setting $C_{t_1}^A = C_{t_1}^{\text{used}} - C^f$.

The period backtracking itself is accomplished by decreasing t_1 by one ($t_1^* = t_1 - 1$) and performing a new iteration of the lot-sizing procedure. In this iteration, the aggregate capacity of the (new) first period t_1^* and of all periods except $t_1^* + 1$ is reset to its original value. The aggregate capacity of period $t_1^* + 1$ ($C_{t_1^*+1}^A$) is set as described above. If, in later iterations, we do one more backtracking from period $t_1^* + 1$ to t_1^*, we take the already reduced capacity $C_{t_1^*+1}^A$ as a basis and reduce it further by the new additionally needed capacity. In this way, we prevent infinite looping as capacities of earlier periods are reduced further and further. In the end, we either find a feasible solution or we backtrack to period 1 and do not find a solution without overtime. If no solution of model HMO without overtime can be found for $t_1 = 1$, the solution procedure quits without a solution—either because there is no feasible solution or because the algorithm is unable to find one.

A formal description of the lot-sizing algorithm with backtracking is presented in Fig. 4.13. It shows the backtracking algorithm incorporated into the lot-sizing procedure as shown in Fig. 4.8. We make use of two additional parameters C_t^{orig} and C_t^{reduced}, representing the original and the reduced aggregate capacity per period. C_t^{reduced} will be updated during the algorithm. Both are initialized with $M \cdot C$, the original aggregate capacity per period.

An example covering three periods is given in Fig. 4.14: The gray bars show the available capacity in each period, the fasciated bars indicate an overtime usage. In iteration 1, t_1 is one and the lot-sizing problem is solved for all periods 1, 2 and 3. Since no overtime is used, t_1 is set to two and the next iteration is executed. In iteration 2, the lot-sizing procedure is not able to find a solution without overtime. For iteration 3, we backtrack t_1 to period 1 and solve an instance with reduced capacity in period 2. A new solution is found and we proceed with $t_1 = 2$. The capacity of period 2 is reset to its original value and iteration 4 finds a feasible solution covering periods 2 and 3. In iteration 5, we observe another overtime usage for $t_1 = 3$ (or a back-order volume after the scheduling procedure), so we backtrack to period 2 and decrease the capacity of period 3 by the overtime amount. In iteration 6, the lot-sizing procedure cannot find a solution without overtime. We set $t_1 = 1$, reset the capacity in period 3 and reduce the capacity in period 2 even further, based on the reduced capacity from iteration 3. We find feasible solutions without overtime in iterations 7, 8 and 9. After iteration 9, we have found a feasible solution to the original problem.

For each period $t_1 = 1, 2, \ldots, T$

1. Set aggregate capacities for HMO:
   Set $C_t^A = \begin{cases} C_t^{\text{orig}} & \text{for } t = t_1 \\ C_t^{\text{reduced}} & \text{for } t = t_1 + 1, t_1 + 2, \ldots, T \end{cases}$
2. Solve HMO covering periods $t_1, t_1 + 1, \ldots, T$
3. Calculate aggregate capacity used in period $t_1 + 1$:
   Set $C_{t_1+1}^{\text{used}} = \sum_{p \in \bar{P}} \left(t_p^u \cdot \left(x_{p, t_1+1} + x_{p, t_1+1}^D \right) + t_p^s \cdot \gamma_{p, t_1+1} \right)$
4. If HMO could be solved without overtime
   - Calculate production quantities and machine assignments for period t_1 (scheduling procedure)
   - If $t_1 = T$ and $b_{pT} > 0$ for any product p
     - Calculate number of units that could be produced using existing setups: Set $n = \sum_{\substack{m \in \bar{M} \\ x_{pTm} > 0}} \left(C_p^c - x_{pTm} - t_p^s \cdot \gamma_{pTm} \right)$
     - Calculate remaining back-order volume:
       Set $b_{pT} = \max\{b_{pT} - n, 0\}$
     - Calculate additionally needed capacity for processing:
       Set $C_p^f = \sum_{p \in \bar{P}} b_{pT} \cdot t_p^u$
     - Calculate additionally needed capacity for setups:
       Set $C_s^f = \sum_{p \in \bar{P}} \left\lceil \frac{b_{pT}}{C_p^s} \right\rceil \cdot t_p^s$
     - Calculate total additionally needed capacity:
       Set $C^f = C_p^f + C_s^f$
     - Reduce aggregate capacity in period t_1:
       Set $C_{t_1}^{\text{reduced}} = \max\left\{C_{t_1}^{\text{used}} - C^f, 0\right\}$
     - Backtrack one period: Set $t_1 = t_1 - 1$
   else
   - if $t_1 > 1$
     - Reduce aggregate capacity in period t_1:
       Set $C_{t_1}^{\text{reduced}} = \max\left\{C_{t_1}^{\text{used}} - C^O, 0\right\}$
     - Reset aggregate capacities in later periods:
       Set $C_t^{\text{reduced}} = C_t^{\text{orig}}$ for all $t = t_1 + 1, t_1 + 2, \ldots, T$
     - Backtrack one period: Set $t_1 = t_1 - 1$
   - else
     - No feasible solution found. Stop.

Fig. 4.13. Lot-sizing procedure with period backtracking

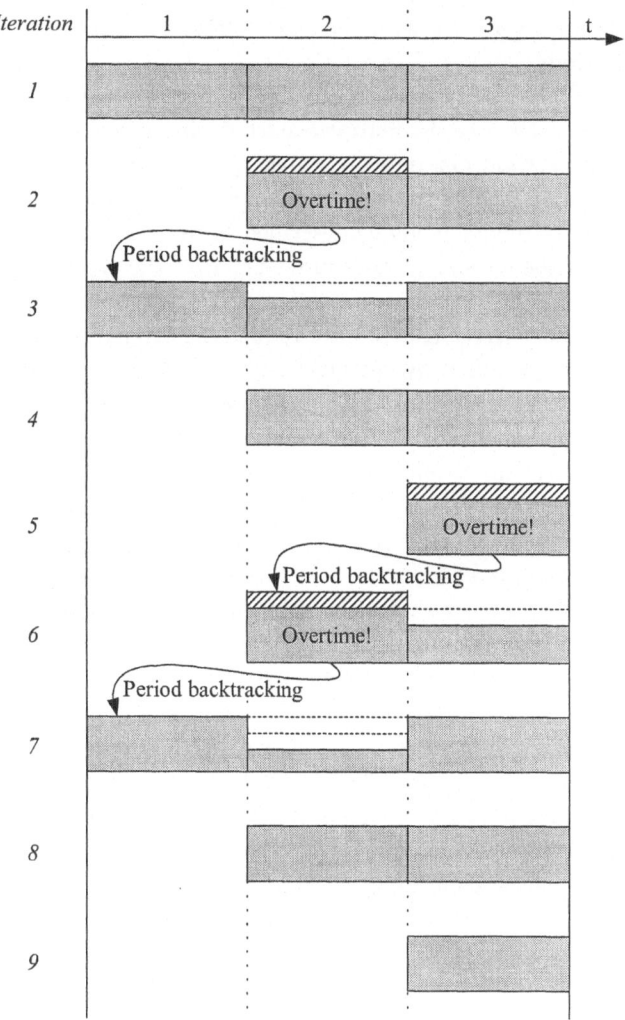

Fig. 4.14. Example of the period backtracking algorithm

4.4.2.4 Capacity Adjustment

When the scheduling procedure is not able to schedule the complete production volume given by a solution of model HMO in any period τ, the latter has overestimated the capacity of this period. There are two possible cases with different consequences: If the production volume was to be put on inventory, the inventory will not be built up. If the produced quantity should satisfy demand of period τ, the volume is back-ordered to period $\tau + 1$. However, the procedure can continue with the next iteration, as was explained in the preceding sections.

In any case, we might be able to find a better solution to the overall problem if no capacity overestimation occurred. Hence, we save the first period when such a misestimation appears. Further, we calculate C^f, the amount of additional capacity needed to produce the complete production volume given by the solution to model HMO as described in Sect. 4.4.2.3 (case 2). After a complete run of the lot-sizing procedure—i.e. after the procedure has found a solution to the original problem or terminated without a solution—we check if a misestimation occurred. If it did, we reduce the original aggregate capacity of the particular period τ by setting $C_\tau^{\text{orig}} = C_\tau^{\text{used}} - C^f$ and perform one more run. We do so until we find a solution without capacity misestimations or until we are no longer able to find a feasible solution. Since we decrease the capacity with each run, infinite looping is impossible.

In the example shown in Fig. 4.14, the lot-sizing procedure cannot find a solution without overtime in iteration 2 (period 2). Since a solution without overtime was found in iteration 1, a capacity overestimation must have occurred in period 1—otherwise the solution of iteration 1 would also represent a solution without overtime for iteration 2. The scheduling procedure has not been able to schedule the given production volume in period 1 and (additional) units had to be back-ordered to period 2. In this example, the capacity adjustment is performed as follows: After finding a feasible solution in iteration 9, we reduce the original aggregate capacity of period 1 by the amount of the misestimation saved in iteration 1 and start another run.

Figure 4.15 shows the complete procedure. For the sake of a tight presentation, step 1 is not split up into all its components as shown in Fig. 4.13.

Repeat

1. Run lot-sizing procedure with period backtracking
   - Save first period τ when the scheduling procedure was not able to schedule the complete production volume given by the solution of model HMO
   - Calculate C^f, the additional capacity needed in τ to schedule the complete production volume given by the solution of model HMO
2. Reduce aggregate capacity in period τ: Set $C_\tau^{\text{orig}} = C_\tau^{\text{orig}} - C^f$
3. Re-initialize all variables except C_t^{orig} for all $t = 1, 2, \ldots, T$

until no capacity misestimation or no solution is found

Fig. 4.15. Lot-sizing procedure with period backtracking and capacity adjustment

4.4.2.5 Procedure Variants

Several variants of the presented procedure are possible. Firstly, we suggested two different objective functions for the scheduling model shown in Sect. 4.4.2.2. One objective function devaluates setup costs by a factor $f \ll 1$, and another one sets $f = 1$, implying that setup costs are not devaluated. In the remainder, we will call the first variant LSP-WSCF (Lot-sizing and Scheduling Procedure With Setup Costs Factor) and the second one LSP-NSCF (Lot-sizing and Scheduling Procedure No Setup Costs Factor). LSP-WSCF tries to schedule the complete production volume given by the lot-sizing procedure. The lot-sizing procedure has an overview over all periods and balances the production volumes accordingly. The scheduling procedure does not have such an overview. Thus, it seems logical for the scheduling procedure to try to stick to the lot-sizing procedure's target production volumes. On the other hand, this might imply setting up a machine for a small number of units at the end of a period and hence high setup costs. If there is leftover capacity in future periods, a better solution might be to postpone the production and save setups. This behavior is adopted by LSP-NSCF. Since LSP-NSCF has a tendency to produce less than the lot-sizing procedure's target production volume, we can expect that more backtracking steps have to be performed to reach feasibility. Therefore, longer run-times of procedure LSP-NSCF can be expected when compared with LSP-WSCF.

Another variant is to combine LSP-WSCF and LSP-NSCF. For this purpose, both variants are used to solve the respective problem and the solution with lower costs is reported. This procedure will be called LSP-BOTH.

The inclusion of the capacity adjustment method (Sect. 4.4.2.4) leads to another type of variants. In the remainder, we denote variants including capacity adjustment by an appendix 'CA'. Variants without this appendix perform only a single run of the lot-sizing procedure (including period backtracking). We can expect faster but worse solutions by these variants.

Together, we have six alternative procedures, namely LSP-WSCF, LSP-NSCF, LSP-BOTH, LSP-WSCF-CA, LSP-NSCF-CA and LSP-BOTH-CA. All of them are included in the computational tests.

4.4.3 Rolling Planning Horizon

In the preceding sections, we assumed a fixed planning horizon of T periods. In practical settings, the lot-sizing and scheduling problem does not appear just once, but repeatedly. A rolling planning horizon is the common solution approach to this situation. With a rolling planning horizon, the T periods contained in the lot-sizing and scheduling model represent the currently upcoming periods of a longer time frame (see e.g. Hax and Candea 1984, p. 76, de Matta and Guignard 1995 or Stadtler 2003). After generating a schedule for T periods, only the production volumes of the first $T^f < T$ periods are fixed and handed over to the shop floor. The range of periods from 1 to T^f is called the frozen interval. Later periods between T^f and T are being rescheduled in a subsequent planning run covering periods $T^f + 1$ to $T^f + T$.

The reason for proceeding this way is as follows: In practical settings, the demands of periods in the distant future are usually more uncertain than demands of close-by periods. At the same time, the plan that is handed over to the shop floor should be based on as accurate and certain data as possible. On the other hand, it is necessary that a schedule also takes the situation of future periods into account to avoid the so-called truncated horizon effect—for example, if a product has to be produced to stock because of a foreseeable lack of capacity in the future. With a rolling planning horizon and periodical replanning, the impact of uncertainty in future periods can be diminished while at the same time the forecasts are incorporated into the planning process. The schedule for future periods is preliminary only, but can—to some extent—be used for further planning steps, such as purchasing of intermediate products.

The proposed lot-sizing and scheduling approach may easily be used in a rolling planning horizon. In fact, such an environment should even speed up computation times. While model HMO will cover all periods of the planning horizon, the lot-sizing and scheduling procedure has to be invoked only for the frozen interval. The schedule for later periods may be read from the solution of model HMO directly, as capacity misestimations in future periods will be tackled in a subsequent planning run. In this way, the number of backtracking steps may be reduced substantially, which further shortens the computation time.

A well-known problem of a rolling planning horizon is its so-called plan nervousness: With subsequent planning intervals, the schedule of periods outside the frozen interval may change substantially. The more changes or the higher the magnitude of the changes, the more nervous the plan. With the proposed planning approach, the level of nervousness can be decided by the planner himself. Decision variables that shall not change in a subsequent planning run are simply fixed to their values from the previous run. This can be done by adding appropriate linear constraints to model HMO. For example, production volumes or setup variables of some products or machines may be fixed, while others remain flexible. It is also possible to define a certain range around the preliminary production volumes by adding lower and upper bound constraints for the respective variables.

To sum up, the planning approach seems well suited for a rolling planning horizon.

4.5 Computational Results

We are interested in solution quality and run-time behavior of the presented heuristics. Due to a lack of applicable problem instances in the literature, test instances have been generated using a tailor-made procedure along general guidelines given by Hooker (1995) and Barr et al. (1995). These authors provide an introduction on how to design and report computational tests of heuristic procedures.

The algorithms are tested with respect to three dimensions. The first set of tests is concerned with different problem characteristics. Small, medium-sized and large test problems are investigated for six factors. A second set of tests examines two internal parameters: The effect of the capacity adjustment routine and the time limit used for solving each model instance. Finally, a last set of tests inspects the impact of problem size, namely the number of products, periods and machines.

The procedures have been implemented using ILOG OPL Studio 3.6 and ILOG OPL Script (see ILOG 2001). Except for the tests investigating the time limit itself, the time limit has been set to 2 minutes for each call of the standard solver (CPLEX 8) to solve model HMO or the scheduling model.

To our knowledge, there is no other non-standard algorithm solving the CLSPL-BOPM. To assess and compare results, we therefore implemented the model CLSPL-BOPM directly in OPL. This model is solved with a time limit, which is adapted to the problem size and mentioned in the respective sections. If the optimal solution cannot be found or verified within the time limit, the best solution found so far is taken as a comparison. In the remainder, we simply call this procedure CLSPL. All tests have been performed on an Intel Pentium IV processor with 2.2 MHz and 1 GB RAM running ILOG OPL Studio 3.6 (CPLEX 8) with standard settings. Interestingly, using the CPLEX preprocessor option to add symmetry cuts to the model had no significant effect, neither on the direct implementation of the CLSPL-BOPM—which contains a lot of symmetric solutions due to the identical parallel machines— nor on the heuristic model.

4.5.1 Effect of Problem Characteristics

4.5.1.1 Experimental Design

The generation of test problems for assessing the effect of the problem characteristics is based on Özdamar and Barbarosoglu (1999). We varied six factors of the problem data to test the performance of the heuristics under different conditions. These factors are prominently used in the lot-sizing literature and tend to have a significant effect on solution quality (e.g. Maes and Van Wassenhove 1986a).

1. The first factor is **demand variation**. In the low variance case, demand figures are generated using a uniform distribution with a minimum value of 90 and a maximum value of 110 ($U(90; 110)$). In the high variance case, a $U(50; 150)$ distribution is used.

2. The second factor is **inventory holding costs**. We set inventory holding costs to $I \cdot U(1; 20)$ for each product individually with two levels of I at 1 (low inventory holding costs) and 5 (high inventory holding costs). Back-order costs are set to 1.25 of the respective product's inventory costs. High inventory costs therefore entail high back-order costs. Hence, the inventory holding costs factor allows a comparison of the heuristics with regard to the relative level of **setup costs**.

3. The third factor is the **capacity utilization** factor CUF. We calculate the average processing time over all machines and periods, APT^{MT}, by

$$APT^{MT} := \sum_{\substack{p \in \bar{P} \\ t \in \bar{T}}} \left(t_p^u d_{pt} \right) / (M \cdot T) \ . \tag{4.75}$$

The period-capacity of a parallel machine is set to $C := APT^{MT}/CUF$ with CUF at levels of 0.85 (high capacity utilization) and 0.7 (low capacity utilization). The resulting machine utilization will be higher than these figures, as setup times will be added. The final utilization is noted in the respective sections.

4. The fourth factor is **setup time**. We calculate the average processing time over all periods caused by an average product, APT^{PT}, by

$$APT^{PT} := \sum_{\substack{p \in \bar{P} \\ t \in \bar{T}}} \left(t_p^u d_{pt} \right) / (P \cdot T) \ . \tag{4.76}$$

(Notice that $APT^{PT} = APT^{MT} \cdot M/P$). For example, an APT^{PT} value of 100 means that an average product needs 100 capacity units for production in each period—on one or more machines. Setup times are generated by using the formula $t_p^s := APT^{PT} \cdot U(1; 3) \cdot S$ for all products $p \in \bar{P}$, where S is held at two levels of 0.05 (low setup times) and 0.25 (high setup times). In other words, in the low setup time case, setup times are between 5% and 15% of the per period processing time of an average product. In the high setup time case, setup times are between 25% and 75% of this figure, yielding very long setup times.

5. The fifth factor is **demand probability**—comparable to the notion of demand lumpiness as introduced by Blackburn and Millen (1980). We hold demand probability at two levels of 100% and 50%, meaning that in the first case, we generate a positive demand figure for all products and

periods according to the demand variation factor above. In the second case, a positive demand is generated every second period on average— and the demand figures are doubled to yield the same average demand. In practice, a random number is drawn from a $U(0;1)$ distribution for each d_{pt}-figure. If the random value is smaller than 0.5, the specific demand value is set to zero. Otherwise, it is set to the double of the value generated according to the demand variation factor.

6. The sixth factor is the **demand pattern**, similar to the one used by Haase (1994). Three different demand patterns are investigated:

   a) Equally distributed, where the demand figures are taken according to the demand variation and demand probability factors as above.

   b) Different sizes, where the products have different demand levels to investigate cases with high and low volume products. For that purpose, a higher demand volume is assigned to products with a higher index number in $p \in \bar{P} = \{1 \dots P\}$. A demand modifier d_p^{mod1} is calculated as

$$d_p^{\mathrm{mod1}} := \frac{2p}{P+1} \qquad\qquad \forall\, p \in \bar{P} . \qquad (4.77)$$

   The demand volume generated according to the demand variation and demand probability factors (denoted by d_{pt}^{old}) is multiplied by this factor, i.e. the final demand volume is set to

$$d_{pt} := d_p^{\mathrm{mod1}} \cdot d_{pt}^{\mathrm{old}} \qquad\qquad \forall\, p \in \bar{P},\, t \in \bar{T} . \qquad (4.78)$$

   Thus, the difference between the demand volume of product p and $p+1$ is $2/(P+1)$-times the demand volume of an average product, which remains unchanged at 100 units.

   c) Positive trend, where the demand volume increases in later periods. A demand modifier d_t^{mod2} is calculated as

$$d_t^{\mathrm{mod2}} := \frac{2t}{T+1} \qquad\qquad \forall\, t \in \bar{T} . \qquad (4.79)$$

   Analogously, the final demand is set to

$$d_{pt} := d_t^{\mathrm{mod2}} \cdot d_{pt}^{\mathrm{old}} \qquad\qquad \forall\, p \in \bar{P},\, t \in \bar{T} . \qquad (4.80)$$

Table 4.2 summarizes the experimental design. In total, there are $2 \times 2 \times 2 \times 2 \times 2 \times 3 = 96$ factor combinations.

The remaining parameters are set as follows: Per unit process times are generated using a $U(5;15)$ distribution, initial back-order and inventory volumes are set to zero and setup costs are drawn from a $U(3\,000; 5\,000)$ distribu-

Table 4.2. Experimental design for the tests of Phase I

Demand variation (DemVar)	low variance: $U(90; 110)$ high variance: $U(50; 150)$
Inventory holding costs (InvCosts)	low inventory holding costs: $1 \cdot U(1; 20)$ high inventory holding costs: $5 \cdot U(1; 20)$
Capacity utilization (CapUtil)	low capacity utilization: $APT^{MT}/0.85$ low capacity utilization: $APT^{MT}/0.7$
Setup time (SetupTime)	low setup times: $APT^{PT} \cdot U(1; 3) \cdot 0.05$ high setup times: $APT^{PT} \cdot U(1; 3) \cdot 0.25$
Demand probability (DemProb)	every period: 100% every second period: 50%
Demand pattern (DemPat)	equal different sizes positive trend

tion. All figures are drawn randomly for each product separately. The initial setup state is also drawn at random for each machine—i.e. each machine is initially set up for a random product at the beginning of the planning horizon.

4.5.1.2 Small Test Problems

To compare the heuristics with an optimal solution, we generated small test problems with 5 products, 4 periods and 4 machines. Ten instances have been created for each of the 96 factor combinations, leading to 960 test problems. 934 of these instances could provably be solved optimally with the direct CLSPL-BOPM implementation and a time limit of 10 minutes. The average run-time of the CLSPL-BOPM was 45 seconds. Its solutions had an average utilization (processing and setup time) of 82%, ranging from 70% to 100% (> 99.89%). The overall results for all procedures are shown in Table 4.3. Results are compared with the best solution over all procedures. Thus, a ratio of one means that the procedure found the best solution. The columns are as follows: the number of instances solved by the procedure (#Inst), the best or minimal ratio achieved over all instances (Min), the worst or maximal ratio (Max), the standard deviation of the ratio (StdDev), the average ratio (Ratio), the number of instances for which the specific procedure found the best solution (#Best) and the average computation time in h:mm:ss (Time). While the ratio calculations include only instances that could be solved by the

Table 4.3. Summary of the results for small test problems (960 instances)

Procedure	#Inst	Min	Max	StdDev	Ratio	#Best	Time
LSP-WSCF	873	1.00	3.39	0.27	1.27	51	0:00:05
LSP-NSCF	856	1.00	2.67	0.27	1.29	46	0:00:07
LSP-BOTH	879	1.00	2.67	0.25	1.26	51	0:00:12
LSP-WSCF-CA	921	1.00	2.24	0.22	1.23	52	0:00:11
LSP-NSCF-CA	893	1.00	2.34	0.22	1.25	47	0:00:25
LSP-BOTH-CA	921	1.00	2.24	0.21	1.22	52	0:00:35
CLSPL	960	1.00	1.00	0.00	1.00	960	0:00:45

specific procedure, the run-times are averaged over all (solved and not solved) instances.

The computation times also include writing of the results to a database and, if applicable, the transformation of input data. The input data is given in the format for the CLSPL-BOPM, so for the other procedures, some parameters (e.g. C_p^c) have to be pre-calculated. Depending on the problem size, the database operations and the transformations need about one second, so they are negligible. Nevertheless, they explain why—in the following sections—the CLSPL uses more time than its designated time limit.

It can be seen that the heuristics have a shorter computation time than the CLSPL. When comparing the run-time of the heuristics, one can see that the capacity adjustment method needs a lot of extra computation time. Therefore, we group the non-CA and the CA procedures together and analyze the groups' run-time separately. A paired t-test comparing the WSCF with the NSCF methods shows that the run-time difference is significant at an error level of 0.5% for both the non-CA and CA groups. LSP-WSCF-CA performs 1.81 runs per instance on average and finds the best solution after 1.46 runs. The respective figures for LSP-NSCF-CA are 2.22 and 1.54. Thus, our theoretical conclusion that the WSCF procedures will be faster than the NSCF procedure (see Sect. 4.4.2.5) is supported by these results. The BOTH procedures obviously have a longer run-time than their respective components. Interestingly, the run-time of LSP-BOTH without capacity adjustment is longer than the one of LSP-WSCF-CA, even though LSP-WSCF-CA leads to better solutions. For all heuristics, the mean computation times of problems that could be solved (see Table 4.4) is shorter than the mean over all problems as reported in Table 4.3.

Neglecting the CLSPL, the best solutions are delivered by LSP-BOTH-CA, shortly followed by LSP-WSCF-CA, but the former needs more than three times the computation time of the latter. In general, the CA procedures lead to significantly better solutions (using a paired t-test on each non-CA

Table 4.4. Computation times for solved instances, small test problems

Procedure	Time
LSP-WSCF	0:00:03
LSP-NSCF	0:00:04
LSP-BOTH	0:00:08
LSP-WSCF-CA	0:00:09
LSP-NSCF-CA	0:00:21
LSP-BOTH-CA	0:00:32
CLSPL	0:00:45

and CA procedure-pair at an error level of 0.5%, only instances solved by both procedures in the comparison). Also within the non-CA and the CA groups, the solution quality differs significantly: ANOVA analyses show that in either group, the BOTH procedures are best at an error level of 5%. Paired t-tests at an error level of 0.5% show that the BOTH procedures perform significantly better than the WSCF ones, which in turn are significantly better than the NSCF ones (taking into account only instances solved by both procedures in the specific comparison).

The procedures using the capacity adjustment method lead to a lower ratio variance. Clearly, if a relatively poor solution is found in the first run, it can be improved in subsequent runs with adjusted capacity. This ultimately leads to a more evenly distributed solution quality. The same holds true for the BOTH variants, as a possibly poor solution of one procedure may be compensated by a better one from the other component. In any case, the latter effect is less strong as can be seen by the rather small difference of the standard deviation of the ratios. An analogous result can be drawn for the number of solved instances. When the specific non-CA (single run) heuristic does not find a solution, a solution may frequently be found in successive runs using capacity adjustment or—for the BOTH methods—using the second procedure.

Tables 4.5 and 4.6 show the performance of the procedures for the different factors. Table 4.7 shows the average run-time of the CLSPL for a comparison. The average ratio is omitted because the procedure found the best solution for all instances. A single asterisk (*) denotes a significant difference in run-time or solution quality at a 5% error level, a double asterisk (**) at an 0.5% error level. Tests have been performed using ANOVA analyses. Again, run-time figures include all instances whereas ratio figures include only the ones solved by the respective procedure.

All heuristics perform significantly better at a low demand probability, with a relatively large difference of the respective means (e.g. 1.16 to 1.38 for LSP-WSCF).

Table 4.5. Factorial design results for small test problems, non-CA procedures

	LSP-WSCF		LSP-NSCF		LSP-BOTH	
	Ratio	Time	Ratio	Time	Ratio	Time
DemVar						
low	1.28	0:00:06*	1.31	0:00:07	1.27	0:00:12
high	1.26	0:00:04	1.28	0:00:08	1.25	0:00:12
InvCosts						
low	1.26	0:00:04*	1.27*	0:00:06*	1.25	0:00:09*
high	1.28	0:00:06	1.31	0:00:09	1.27	0:00:14
CapUtil						
low	1.25*	0:00:01**	1.27**	0:00:02**	1.24**	0:00:03**
high	1.29	0:00:09	1.33	0:00:12	1.29	0:00:21
SetupTime						
low	1.25*	0:00:01**	1.28	0:00:02**	1.24*	0:00:04**
high	1.29	0:00:08	1.31	0:00:12	1.28	0:00:20
DemProb						
every	1.38**	0:00:07**	1.40**	0:00:10**	1.37**	0:00:16**
every 2nd	1.16	0:00:03	1.18	0:00:05	1.15	0:00:07
DemPat						
equal	1.33**	0:00:05	1.34**	0:00:07	1.32**	0:00:13
diff sizes	1.19	0:00:04	1.23	0:00:06	1.18	0:00:10
pos trend	1.29	0:00:05	1.31	0:00:08	1.28	0:00:14

Another important factor for solution quality is the demand pattern. All procedures perform significantly better when the demand volume is unevenly distributed among products (different sizes case). This is not surprising when we reconsider the underlying idea of the lot-sizing and scheduling model. It is based on a situation of a mix of high and low volume products (see Sect. 4.4.1.1). Compared with the equally distributed demand volume pattern, the procedures also perform better when there is a positive trend in the demand volume. Interestingly, for the non-CA procedures, there seems to be a relatively large difference between the ratios of the 'different sizes' and the 'positive trend' case (e.g. 1.19 to 1.29 for LSP-WSCF), whereas for the CA procedures, the greater difference lies between the positive trend and the equally distributed case (e.g. 1.22 to 1.29 for LSP-WSCF-CA).

The heuristics perform significantly better under low capacity utilization, but the difference of the means is rather moderate (e.g. 1.25 to 1.29 for LSP-WSCF). The factor measures capacity utilization without setup times. Within the low and high capacity utilization groups, there are differences in utiliza-

Table 4.6. Factorial design results for small test problems, CA procedures

	LSP-WSCF-CA		LSP-NSCF-CA		LSP-BOTH-CA	
	Ratio	Time	Ratio	Time	Ratio	Time
DemVar						
low	1.25*	0:00:13	1.27*	0:00:29	1.24*	0:00:41
high	1.21	0:00:08	1.23	0:00:20	1.20	0:00:28
InvCosts						
low	1.22	0:00:07*	1.24	0:00:13*	1.22	0:00:20*
high	1.23	0:00:14	1.26	0:00:36	1.22	0:00:50
CapUtil						
low	1.20**	0:00:04**	1.22**	0:00:07**	1.20**	0:00:11**
high	1.26	0:00:17	1.29	0:00:42	1.25	0:00:59
SetupTime						
low	1.22	0:00:03**	1.25	0:00:20	1.22	0:00:23*
high	1.23	0:00:19	1.26	0:00:29	1.23	0:00:47
DemProb						
every	1.32**	0:00:16**	1.35**	0:00:37*	1.31**	0:00:53**
every 2nd	1.14	0:00:05	1.15	0:00:12	1.13	0:00:17
DemPat						
equal	1.29**	0:00:12	1.30**	0:00:19	1.28**	0:00:31
diff sizes	1.17	0:00:08	1.20	0:00:31	1.17	0:00:39
pos trend	1.22	0:00:12	1.24	0:00:23	1.21	0:00:35

tion including setup times. Further ANOVA tests have shown that even within these groups, the problems with low utilization including setups lead to significantly better results.

The level of setup times seems to have a small effect on solution quality—in favor of a low setup time—but the effect is only significant for two out of three non-CA procedures. The situation is opposite for demand variation. Here, the effect is only significant for (all three) CA procedures. Just as the procedures perform better at a low demand probability, they also perform better at a high demand variance, which—from a conceptual point of view—is a similar situation.

Inventory holding costs are a significant factor for LSP-NSCF only. In any case, all procedures generate slightly better results for the low level case on average.

The relative ranking of the procedures as depicted in Table 4.3 remains the same for the different factors. Only for an unevenly distributed demand over the products (the 'different sizes' demand pattern), the average solution qual-

Table 4.7. Factorial design computation times for small test problems, CLSPL

	CLSPL Time
DemVar	
low	0:00:56**
high	0:00:33
InvCosts	
low	0:00:10**
high	0:01:19
CapUtil	
low	0:00:21**
high	0:01:08
SetupTime	
low	0:00:22**
high	0:01:07
DemProb	
every	0:01:13**
every 2nd	0:00:16
DemPat	
equal	0:00:57*
diff sizes	0:00:31
pos trend	0:00:46

ity of LSP-WSCF is better than the one of LSP-NSCF-CA, which is in contrast to the overall ranking. The difference is not statistically significant, however.

Regarding the computation times, we can say that for all LSP heuristics—with two exceptions—inventory costs, capacity utilization, setup times and demand probability have a significant effect. The effect is stronger for the latter three factors than for inventory costs. For the CLSPL, all factors have a significant effect on computation time. The classes of problems that yield longer computation times for the heuristics do so also for the CLSPL. It seems these are the more difficult problems in general. Most interestingly, with the heuristics, shorter computation times are achieved for classes that lead to better solutions. This also holds true for most of the statistically insignificant results.

Table 4.8 shows how many times a procedure found the best solution for all instances and the number of optimally solved instances by the CLSPL. In general, the figures support the above conclusions. What is remarkable is the relatively high number of best solutions for the low capacity utilization class

Table 4.8. Number of best solutions and optimally solved instances for small test problems (960 instances)

	LSP WSCF	LSP NSCF	LSP BOTH	LSP WSCF CA	LSP NSCF CA	LSP BOTH CA	CLSPL	CLSPL
	#Best	#Best	#Best	#Best	#Best	#Best	#Best	#Opt
DemVar								
low	27	25	27	28	26	28	480	461
high	24	21	24	24	21	24	480	473
InvCosts								
low	27	24	27	27	24	27	480	480
high	24	22	24	25	23	25	480	454
CapUtil								
low	39	37	39	39	37	39	480	478
high	12	9	12	13	10	13	480	456
SetupTime								
low	24	22	24	24	22	24	480	476
high	27	24	27	28	25	28	480	458
DemProb								
every	8	6	8	8	6	8	480	454
every 2nd	43	40	43	44	41	44	480	480
DemPat								
equal	17	17	17	18	18	18	320	307
diff sizes	12	11	12	12	11	12	320	314
pos trend	22	18	22	22	18	22	320	313
Total	51	46	51	52	47	52	960	934

in comparison with the high capacity utilization class. While Tables 4.5 and 4.6 also indicated a better performance under low capacity utilization, the effect was comparably smaller. Moreover, for the setup time and the demand pattern factor, more best solutions could be found for problem classes that had a worse ratio in Tables 4.5 and 4.6.

Let's summarize the results for the small test problems: The six factors have a similar influence on all procedure variants. The heuristics are better and faster under low demand probability, an uneven distribution of demand over products (and periods), and low capacity utilization. With respect to their computation times, heuristics LSP-WSCF and LSP-WSCF-CA seem to perform best.

4.5.1.3 Medium-Sized Test Problems

The medium-sized problems consist of instances with 15 products, 4 periods and 6 machines. Five instances have been created for each of the 96 factor combinations, leading to 480 test problems. Using a time limit of 10 minutes, only 284 instances could be solved with the direct CLSPL-BOPM implementation, and only two of them provably to optimality. For the other instances, the CLSPL procedure did not find any solution. The average utilization including setup times was 81%, ranging from 70% to 99% (measured by the solutions of LSP-BOTH-CA, because this procedure solved the highest number of instances). The overall results for all procedures are presented in Table 4.9. Again, the results are compared with the best solution over all procedures and the columns show the same figures as for the small test problems. The ratio calculations include only instances that could be solved by the specific procedure, the run-times are averaged over all (solved and not solved) instances. For one instance, LSP-NSCF-CA did not terminate within 24 hours because the capacity adjustment between runs was very small and did not lead to substantially different solutions. Even though a solution has been found in each of its runs, we disregard the instance for this procedure. For procedure LSP-BOTH-CA, we disregard the instance when reporting computation times but consider it solved and use the solution of LSP-WSCF-CA. In any case, the effect is almost completely averaged out by the other instances.

29 instances could not be solved by any procedure. The generation of test instances does not guarantee that a problem has a feasible solution. Since some other solutions showed a utilization of 99%, it seems likely that at least some of the 29 problems do not have a feasible solution at all.

We can see that the NSCF procedures have a substantially longer run-time than the respective WSCF ones. For both the CA and the non-CA procedure pairs, this observation is confirmed by paired t-tests at a significance level of 0.5%. The LSP-NSCF-CA—and hence, the LSP-BOTH-CA as well—has a longer run-time than the CLSPL, which has been terminated after a time limit of 10 minutes. Just as for the small test problems, the LSP-BOTH runs longer than the LSP-WSCF-CA but delivers worse solutions on average. The LSP-WSCF-CA performs 2.58 runs per instance on average and finds the best solution after 1.68 runs. The respective figures for LSP-NSCF-CA are 5.34 and 1.86. These figures are higher than for the small test problems, which can be explained by the higher number of products. The more products have to be scheduled, the higher the probability that model HMO may not correctly estimate the capacity. Moreover, step 3 of the scheduling procedure becomes more difficult: Even though the complete production volume as given by a solution of model HMO could be assigned in the actual period, the scheduling procedure might not find such a feasible schedule. Eventually, a larger number

Table 4.9. Summary of the results for medium-sized test problems (480 instances)

Procedure	#Inst	Min	Max	StdDev	Ratio	#Best	Time
LSP-WSCF	441	1.00	1.60	0.09	1.07	156	0:03:05
LSP-NSCF	411	1.00	1.70	0.11	1.08	87	0:05:14
LSP-BOTH	442	1.00	1.60	0.09	1.05	189	0:08:19
LSP-WSCF-CA	450	1.00	1.54	0.07	1.03	262	0:07:48
LSP-NSCF-CA	416	1.00	1.54	0.09	1.06	153	0:20:41
LSP-BOTH-CA	451	1.00	1.54	0.06	1.02	340	0:28:29
CLSPL	284	1.00	2.36	0.18	1.10	112	0:10:00

Table 4.10. Computation times for solved instances, medium-sized test problems

Procedure	Time
LSP-WSCF	0:02:21
LSP-NSCF	0:03:13
LSP-BOTH	0:07:11
LSP-WSCF-CA	0:07:09
LSP-NSCF-CA	0:19:58
LSP-BOTH-CA	0:28:22
CLSPL	0:09:59

of capacity adjustments (i.e. runs) has to be performed. As already seen for the small test problems and in line with our theoretical reasoning, the figures for LSP-NSCF-CA are higher than for LSP-WSCF-CA. The difference is greater than for the small test problems, which also explains the substantial difference in the run-times of the two procedures.

Table 4.10 shows the mean computation times for instances that could be solved by the specific procedures, which are again shorter than the mean times over all (solved and not-solved) instances as reported in Table 4.9.

Balancing solution quality and computation times, LSP-WSCF-CA performs best. Combined with the LSP-NSCF-CA procedure, the LSP-BOTH-CA is able to improve the solution quality slightly, but the run-times are quadrupled. On average, all heuristics deliver better solutions than the direct CLSPL-BOPM implementation. This effect is significant at an 0.5% error level (paired t-tests, only instances solved by both procedures in the specific comparison) for all heuristics except for LSP-WSCF, where it is still significant at a 5% error level and LSP-NSCF, where it is not statistically significant.

Using the same test, the three sets of CA procedures are confirmed to perform better than their non-CA counterparts. The WSCF variants lead to

Table 4.11. Factorial design results for medium-sized test problems, non-CA procedures

	LSP-WSCF		LSP-NSCF		LSP-BOTH	
	Ratio	Time	Ratio	Time	Ratio	Time
DemVar						
low	1.07	0:02:50	1.09	0:04:51	1.05	0:07:40
high	1.06	0:03:20	1.08	0:05:37	1.05	0:08:57
InvCosts						
low	1.08*	0:02:32*	1.08	0:04:01*	1.06*	0:06:34*
high	1.06	0:03:37	1.08	0:06:27	1.04	0:10:04
CapUtil						
low	1.06*	0:00:23**	1.07*	0:00:28**	1.05	0:00:50**
high	1.08	0:05:47	1.10	0:10:01	1.05	0:15:47
SetupTime						
low	1.06	0:01:05**	1.09	0:01:55**	1.05	0:03:00**
high	1.07	0:05:04	1.08	0:08:33	1.05	0:13:38
DemProb						
every	1.08**	0:03:12	1.10**	0:05:48	1.07**	0:09:00
every 2nd	1.05	0:02:57	1.07	0:04:40	1.04	0:07:37
DemPat						
equal	1.07**	0:03:36	1.09*	0:05:35	1.05**	0:09:12
diff sizes	1.04	0:02:37	1.06	0:05:28	1.03	0:08:05
pos trend	1.09	0:03:00	1.10	0:04:39	1.08	0:07:39

better results than the NSCF ones, too. Not surprisingly, the BOTH variants deliver the best solutions (all results are significant at an error level of 0.5%).

Concerning the cost ratio variance and the number of solved instances, the findings from the small test problems are supported by the medium sized problems. However, the effects are less strong. It should be mentioned that all heuristics are able to find solutions to a fundamentally greater number of instances than the direct CLSPL-BOPM implementation.

The results of the factorial analysis are shown in Tables 4.11 and 4.12 in the same way as for the small test problems. Run-time figures include all instances, ratio figures include only the ones solved by the respective procedure.

While the general picture is similar to the small test problems, there are some apparent differences. The most striking is that LSP-NSCF-CA does not perform significantly better or worse for any of the six factors. All its results are on average between 5% and 6% worse than the best known solutions.

Table 4.12. Factorial design results for medium-sized test problems, CA procedures

	LSP-WSCF-CA		LSP-NSCF-CA		LSP-BOTH-CA	
	Ratio	Time	Ratio	Time	Ratio	Time
DemVar						
low	1.04	0:07:18	1.06	0:18:00	1.03	0:25:18
high	1.03	0:08:17	1.05	0:23:23	1.02	0:31:34
InvCosts						
low	1.05**	0:05:31**	1.06	0:12:36**	1.04**	0:18:07**
high	1.02	0:10:04	1.05	0:28:48	1.01	0:38:45
CapUtil						
low	1.03	0:01:12**	1.05	0:04:58**	1.03	0:06:09**
high	1.03	0:14:23	1.06	0:36:20	1.02	0:50:43
SetupTime						
low	1.04*	0:02:49**	1.06	0:23:20	1.03*	0:26:09
high	1.02	0:12:46	1.05	0:18:02	1.02	0:30:43
DemProb						
every	1.04*	0:08:23	1.06	0:22:00	1.03*	0:30:17
every 2nd	1.02	0:07:12	1.05	0:19:23	1.02	0:26:35
DemPat						
equal	1.04	0:08:07	1.06	0:18:04	1.03*	0:26:11
diff sizes	1.02	0:07:28	1.05	0:21:21	1.01	0:28:41
pos trend	1.04	0:07:47	1.06	0:22:39	1.03	0:30:26

A low demand probability still improves solution quality, but the difference in the respective means is smaller than for the small problems. The results for the demand variation factor are similar. As mentioned before, both factors involve the stability of the demand volume. A high demand variance seems to improve solution quality slightly, but this effect is not significant for any of the six heuristics.

An unevenly distributed demand among products remains the most advantageous demand pattern, but for the CA procedures, this effect is not significant (except for LSP-BOTH-CA at an error level of 5%). The level of setup times does not seem to have a systematical effect on the cost ratio.

The capacity utilization factor had a moderate but significant effect on the cost ratio for the small test problems. For the medium-sized problems, it is even less strong and only significant for LSP-WSCF and LSP-NSCF. Surprisingly, LSP-BOTH-CA performs better for highly capacitated problems while its two components, LSP-WSCF-CA and LSP-NSCF-CA, perform equally well, respectively worse, for these problem sets.

Table 4.13. Factorial design results for medium-sized test problems, CLSPL

	CLSPL	
	Ratio	**Time**
DemVar		
low	1.10	0:10:01
high	1.11	0:09:58
InvCosts		
low	1.07**	0:09:58
high	1.15	0:10:01
CapUtil		
low	1.09	0:09:58
high	1.13	0:10:01
SetupTime		
low	1.09	0:09:58
high	1.12	0:10:01
DemProb		
every	1.12	0:10:01
every 2nd	1.09	0:09:59
DemPat		
equal	1.12*	0:09:57
diff sizes	1.12	0:10:01
pos trend	1.07	0:10:01

The results for the inventory holding costs are contrary to the small test problems. For the medium-sized problems, high inventory costs lead to better solutions than low inventory holding costs. The effect is significant for all procedures except for the NSCF variants. To some extent, it might be explained by the poor performance of the CLSPL, which serves as a comparison. Its results are detailed in Table 4.13. As can be seen, the CLSPL performs worse for high inventory costs (and also for highly capacitated problems, which—even though not statistically significant—might explain the other procedures' better performance for these instances).

The relative ranking of the heuristics as shown in Table 4.9 remains the same for the different factors—with the exception of the 'different sizes' demand pattern. Analogously to the small test problems, the LSP-WSCF performs better than the LSP-NSCF-CA for this problem set. However, also for the medium sized problems, the difference is not statistically significant.

The results for the computation times are very similar to the small test problems: Inventory costs, capacity utilization and setup times have a statistically significant effect on computation times. Inventory costs and capacity utilization have a statistically significant effect on computation times on all six variants, setup times on four of them. As already noticed for the small test problems, a high capacity utilization and high inventory costs require longer computation times for all heuristics. The differences to the small test problems are as follows: A low demand probability still leads to faster solutions, but the effect is less strong and not significant for any of the procedures. A high demand variance leads to (not statistically significant) longer computation times for the medium-sized test problems, contrarily to the small test problems. It is interesting to mention that, while long setup times have a strong and (except for the LSP-BOTH-CA) significant prolonging effect for the other procedures, the LSP-NSCF-CA works faster for problems with high setup times. Nevertheless, this effect is not statistically significant. The runtime for some problem sets is relatively high. We elaborate on that at the end of this section.

Table 4.14 lists the number of best solutions for each procedure as well as the number of optimally solved instances by the CLSPL. The findings are in line with what has been reported for the cost ratio. In contrast to the small test problems, the heuristics find more best solutions for high inventory costs than for low inventory costs. It is remarkable that—except for the LSP-NSCF—more best solutions are found for a high demand probability, while the cost ratio is significantly better with a low demand probability.

Finally, Table 4.15 reports on the number of instances that could not be solved by the individual procedures. Besides the fact that all heuristics solve substantially more instances than the direct CLSPL-BOPM implementation, it can be observed that the NSCF variants solve fewer instances than the other heuristics. In general, it seems that problems with a high capacity utilization and high setup times are more difficult to solve. As mentioned above, it is not clear whether all of these instances do have feasible solutions or not. In addition, the CLSPL had difficulties with high inventory costs and a high demand probability.

The factors have a strong effect on the number of instances that can be solved by a procedure as well as on the computation times. Problems that combine high inventory costs with a high capacity utilization—the two factors that have a significant influence on the run-time of all variants—need a much higher computation time than the other problems. Except for LSP-NSCF-CA and LSP-BOTH-CA, setup times also have a significantly strong effect on computation time. The average computation times for problems with high inventory costs and high capacity utilization as well as for problems with high inventory costs, high capacity utilization and high setup times are noted in Table 4.16 and 4.17. As usual, the time-figures include all (120 respectively 60)

Table 4.14. Number of best solutions and optimally solved instances for medium-sized test problems (480 instances)

	LSP WSCF	LSP NSCF	LSP BOTH	LSP WSCF CA	LSP NSCF CA	LSP BOTH CA	CLSPL	CLSPL
	#Best	#Best	#Best	#Best	#Best	#Best	#Best	#Opt
DemVar								
low	83	46	100	132	79	170	57	0
high	73	41	89	130	74	170	55	2
InvCosts								
low	77	36	89	115	58	143	87	2
high	79	51	100	147	95	197	25	0
CapUtil								
low	99	66	114	144	95	170	71	2
high	57	21	75	118	58	170	41	0
SetupTime								
low	79	39	91	121	77	161	79	2
high	77	48	98	141	76	179	33	0
DemProb								
every	81	40	96	145	78	184	43	1
every 2nd	75	47	93	117	75	156	69	1
DemPat								
equal	54	38	66	87	57	113	40	2
diff sizes	69	24	84	91	40	122	28	0
pos trend	33	25	39	84	56	105	44	0
Total	156	87	189	262	153	340	112	2

instances—the ones that could be solved as well as the ones that could not be solved by the respective procedures. All columns are analogous to Table 4.9.

It can be seen that some variants outreach the 10 minute time limit that has originally been set for the CLSPL. To get a more accurate comparison, we let the CLSPL solve the instances again, with a time limit of 60 minutes. The results are also depicted in Tables 4.16 and 4.17 (CLSPL 60 min).

Not surprisingly, with a longer time limit, the results of the CLSPL become more competitive. However, the LSP-WSCF-CA reaches a very similar—in the case of high setup times even better—solution quality at one third respectively half of the computation time. The differences in solution quality between LSP-WSCF-CA and CLSPL 60 min are not statistically significant.

Table 4.15. Number of instances not solved for medium-sized test problems (480 instances)

	LSP WSCF	LSP NSCF	LSP BOTH	LSP WSCF CA	LSP NSCF CA	LSP BOTH CA	CLSPL
	#Inst	#Inst	#Inst	#Inst	#Inst	#Inst	#Inst
DemVar							
low	22	35	22	14	32	14	94
high	17	34	16	16	32	15	102
InvCosts							
low	15	32	15	11	28	11	61
high	24	37	23	19	36	18	135
CapUtil							
low	0	0	0	0	1	0	53
high	39	69	38	30	63	29	143
SetupTime							
low	0	0	0	0	0	0	48
high	39	69	38	30	64	29	148
DemProb							
every	17	29	16	15	28	14	127
every 2nd	22	40	22	15	36	15	69
DemPat							
equal	11	20	11	7	17	7	62
diff sizes	10	24	10	10	24	10	75
pos trend	18	25	17	13	23	12	59
Total	39	69	38	30	64	29	196

Moreover, there are relatively many instances that cannot be solved by the CLSPL with a 60 minute time limit. Though the number of solvable instances increases compared with the 10 minute time limit, the other procedures solve substantially more instances. The figures for the high setup times show this effect very clearly: While the CLSPL with a 10 minute time limit solved only a single instance, the CLSPL with a 60 minute time limit can solve 9 instances. On the other hand, the LSP-WSCF-CA solves 41 instances in less than 29 minutes on average.

The NSCF variants especially lead to long computation times. We have thus tested two more variants which are slight variations of LSP-NSCF and LSP-NSCF-CA: Whenever a capacity reduction (during a period backtrack-

Table 4.16. Summary of the results for medium-sized test problems with high inventory costs and high capacity utilization (120 instances)

Procedure	#Inst	Min	Max	StdDev	Ratio	#Best	Time
LSP-WSCF	96	1.00	1.33	0.08	1.079	20	0:06:53
LSP-NSCF	83	1.00	1.54	0.12	1.125	10	0:12:27
LSP-BOTH	97	1.00	1.32	0.07	1.062	29	0:19:20
LSP-WSCF-CA	101	1.00	1.26	0.06	1.037	45	0:18:24
LSP-NSCF-CA	85	1.00	1.54	0.09	1.080	19	0:50:50
LSP-BOTH-CA	102	1.00	1.22	0.05	1.025	63	1:09:15
CLSPL 10 min	32	1.00	1.57	0.18	1.165	9	0:10:01
CLSPL 60 min	68	1.00	1.42	0.07	1.034	41	1:00:01

Table 4.17. Summary of the results for medium-sized test problems with high inventory costs, high capacity utilization and high setup times (60 instances)

Procedure	#Inst	Min	Max	StdDev	Ratio	#Best	Time
LSP-WSCF	36	1.00	1.27	0.08	1.059	16	0:11:15
LSP-NSCF	23	1.00	1.29	0.08	1.073	7	0:20:01
LSP-BOTH	37	1.00	1.27	0.06	1.029	23	0:31:17
LSP-WSCF-CA	41	1.00	1.20	0.04	1.019	29	0:28:48
LSP-NSCF-CA	25	1.00	1.29	0.07	1.061	9	0:34:57
LSP-BOTH-CA	42	1.00	1.08	0.01	1.004	38	1:03:45
CLSPL 10 min	1	1.00	1.00	-	1.000	1	0:10:01
CLSPL 60 min	9	1.00	1.18	0.06	1.023	6	1:00:01

ing) or a capacity adjustment is performed, the aggregate capacity is reduced or adjusted by at least 20% of the actual value. The new variants are called LSP-NSCF-20% and LSP-NSCF-CA-20%, respectively. By reducing (and adjusting) the capacity by at least 20%, we expect shorter run-times as consecutive backtracking iterations might be combined. On the other hand, we expect worse solutions because a reduction of 20% might skip the optimal value. It might also lead to infeasible capacity restrictions, which should decrease the number of solvable instances. The summarized results for all 480 medium-sized test problems are presented in Table 4.18. The BOTH procedures combine the regular WSCF procedures with their respective NSCF-20% counterparts. The values for the WSCF procedures are slightly different from the ones listed in Table 4.9, because in some cases, the new 20%-variants have been able to find better solutions than the other procedures.

The results affirm our theoretical reasoning. A capacity reduction/adjustment of at least 20% speeds up the NSCF variants: LSP-NSCF-20% needs

Table 4.18. Summary of the results for medium-sized test problems for NSCF variants with minimum capacity reduction/adjustment (480 instances)

Procedure	#Inst	Min	Max	StdDev	Ratio	#Best	Time
LSP-WSCF	441	1.00	1.60	0.10	1.07	152	0:03:05
LSP-NSCF-20%	363	1.00	1.72	0.13	1.11	70	0:02:36
LSP-BOTH	441	1.00	1.60	0.08	1.05	176	0:05:41
LSP-WSCF-CA	450	1.00	1.54	0.07	1.03	252	0:07:48
LSP-NSCF-CA-20%	366	1.00	1.65	0.11	1.09	77	0:07:26
LSP-BOTH-CA	450	1.00	1.42	0.06	1.03	282	0:15:14
CLSPL	284	1.00	2.36	0.18	1.10	110	0:10:00

about 50% of the run-time of LSP-NSCF, LSP-NSCF-CA-20% only needs about 36% of the run-time of LSP-NSCF-CA. LSP-NSCF-20% and LSP-NSCF-CA-20% are slightly faster than the unmodified versions of LSP-WSCF and LSP-WSCF-CA, respectively. For the non-CA procedures, this difference is significant at a 5% error level. However, the average solution quality deteriorates. The 20% variants are about 3 percentage points worse than the unmodified NSCF versions (see Table 4.9). As anticipated, the number of solvable instances also decreases. About 12% fewer instances can be solved. Problems with high inventory costs, high capacity utilization and high setup times are especially problematic. Those problems have already been identified as the most difficult ones before. Both 20%-procedures can solve only 3 of the 60 instances. Their unmodified counterparts solve 23 respectively 25 problems (see Table 4.17).

As a summary for the medium-sized problems, the effects of the six factors are less systematic than for the small test problems. In general, the heuristics perform better for high inventory costs and low demand probability. They are much faster for low inventory costs, low capacity utilization and—except for LSP-NSCF-CA—for low setup times. Again, the heuristics LSP-WSCF and LSP-WSCF-CA seem to perform best with respect to their computation times.

4.5.1.4 Large Test Problems

The large test problems comprise 288 instances with 20 products, 6 periods and 10 machines. Three instances have been created for each of the 96 factor combinations. Only 97 instances could be solved with the direct CLSPL implementation, even though its time limit was prolonged to 60 minutes to account for the problem size. No instance could provably be solved optimally. The capacity adjustment (CA) heuristics have been excluded from the analysis because of their relatively long computation times.

Table 4.19. Summary of the results for large test problems (288 instances)

Procedure	#Inst	Min	Max	StdDev	Ratio	#Best	Time
LSP-WSCF	285	1.00	1.28	0.04	1.02	185	0:24:56
LSP-NSCF	264	1.00	1.46	0.09	1.07	85	1:07:06
LSP-BOTH	285	1.00	1.28	0.03	1.00	269	1:31:52
CLSPL	97	1.00	4.33	0.41	1.26	16	1:00:02

The overall results are presented in Table 4.19. Again, the results are compared with the best solution over all tested procedures and the columns are the same as for the small and medium-sized test problems. The ratio calculations include only instances that could be solved by the specific procedure, the run-times are averaged over all (solved and not solved) instances. Two instances could not be solved within 24 hours by the LSP-NSCF. The capacity reduction was not large enough to lead to substantially different solutions, and the heuristic kept on moving forward and backtracking. For one of these instances, the LSP-NSCF terminated after about 38 hours without a solution. For the other instance, a solution was found after about 105 hours. In order not to distort the analyses, we treat these two instances as outliers: They are not considered when reporting run-times of the LSP-NSCF and LSP-BOTH.

The average utilization including setup times over all instances was 83%, ranging from 71% to 99% (measured by the solutions of LSP-BOTH). The generation of test instances does not guarantee that a problem has a feasible solution. Thus, we cannot say whether a feasible solution exists for any of the three instances that cannot be solved by LSP-WSCF.

Confirming our earlier findings, LSP-NSCF has a substantially longer run-time than LSP-WSCF. Paired t-tests at an error level of 0.5% show that this difference is significant. The LSP-WSCF is also significantly faster than the CLSPL with a 60 minute time limit. The two additional seconds of its run-time in Table 4.19 originate from database operations. The different average computation times of the LSP-NSCF and the CLSPL are not statistically significant at an error level of 5%. Including the two above-mentioned outliers, the average LSP-NSCF run-time is 1:36:36. However, the difference to the average CLSPL run-time remains statistically insignificant.

Table 4.20 shows the mean computation times for instances that could be solved by the specific procedures. LSP-WSCF and LSP-BOTH have been able to solve almost every instance, and thus their figures are very similar to the ones reported in Table 4.19. Analogous to the small and medium-sized test problems, the LSP-NSCF has a lower average run-time for the instances that it is able to solve than for all—solved and not-solved—instances.

Table 4.20. Computation times for solved instances, large test problems

Procedure	Time
LSP-WSCF	0:25:03
LSP-NSCF	0:51:15
LSP-BOTH	1:32:39
CLSPL	1:00:02

Balancing solution quality and computation time, LSP-WSCF performs best. LSP-BOTH is able to find slightly better solutions, but the computation time incurred by its LSP-NSCF component is much higher. The solutions of the direct CLSPL implementation are substantially worse—and far fewer instances could be solved. The average ratios (solution quality) for all procedures are significantly different at an 0.5% error level (paired t-tests over instances solved by both procedures in the comparison for all procedure pairs as well as an ANOVA analysis for all three heuristic variants).

The results for the different factors are shown in Table 4.21. As for the small and medium-sized problems, run-time figures include all instances, ratio figures include only the ones solved by the respective procedure. Table 4.22 shows the analogous results for the direct CLSPL implementation.

The solution quality of the LSP-WSCF is only significantly affected by the level of inventory costs. For the LSP-NSCF, demand variation and capacity utilization also have an influence. The solution quality of the combined LSP-BOTH is influenced by all factors except demand variation. But all these effects are relatively small as measured by the absolute differences of the average ratios. LSP-BOTH always leads to solutions between 0% and 1% from the best known solution, LSP-WSCF between 1% and 2%. This is due to the fact that these two procedures found the vast majority of best solutions over all factors (Table 4.23). The average ratios of solutions provided by LSP-NSCF are between 6% and 9% worse than the best known solutions. The CLSPL solutions are substantially worse for all factors, with the demand pattern and demand probability factors having a statistically significant effect. The highest deviation from the best known solutions are obtained for problems with a positive demand in every period. The CLSPL-solutions for these problems are on average 137% worse than the best ones. As shown in Table 4.24, only four of these 144 instances could be solved at all (the four ratios are 1.27, 1.34, 2.53 and 4.33).

The relative ranking of the heuristics as shown in Table 4.19 remains the same for all factors.

Regarding computation times, we can see that the capacity utilization and the setup time factor have a statistically significant effect on all variants—at

Table 4.21. Factorial design results for large test problems

	LSP-WSCF		LSP-NSCF		LSP-BOTH	
	Ratio	Time	Ratio	Time	Ratio	Time
DemVar						
low	1.02	0:26:11	1.06*	1:11:19	1.01	1:36:56
high	1.02	0:23:41	1.08	1:02:52	1.00	1:26:46
InvCosts						
low	1.02*	0:16:45*	1.06*	1:17:33	1.01*	1:34:25
high	1.01	0:33:07	1.08	0:56:45	1.00	1:29:21
CapUtil						
low	1.02	0:10:52**	1.08*	0:20:05**	1.01*	0:30:57**
high	1.02	0:38:59	1.06	1:55:08	1.00	2:34:05
SetupTime						
low	1.02	0:13:22**	1.07	0:39:18**	1.01*	0:52:05**
high	1.02	0:36:30	1.07	1:34:43	1.00	2:11:23
DemProb						
every	1.02	0:33:57*	1.06	1:01:29	1.00*	1:35:36
every 2nd	1.02	0:15:55	1.08	1:12:46	1.01	1:28:07
DemPat						
equal	1.02	0:21:51*	1.06	0:50:11	1.00*	1:12:02
diff sizes	1.01	0:14:26	1.06	1:24:28	1.00	1:38:55
pos trend	1.02	0:38:31	1.09	1:06:50	1.01	1:45:00

an error level of 0.5%. A high capacity utilization as well as high setup times lead to longer run-times. The run-time of LSP-WSCF is further affected by all other factors except demand variance, but these effects are only significant at a 5% error level. Compared with the small and medium-sized problems, we find that inventory costs are not a significant factor on the run-time of LSP-NSCF and LSP-BOTH any more. For large problems, the LSP-NSCF needs even longer for problems with low inventory costs, while the LSP-WSCF retains a (significantly) longer run-time for high inventory costs. For the first time, the demand pattern factor shows a significant effect on a procedure's run-time: An uneven demand distribution over products leads to relatively short computation times, and a positive trend provokes the longest average run-times for the LSP-WSCF. Although not statistically significant, we can see that the LSP-NSCF behaves differently: Its run-times are shortest for an equally distributed demand pattern and longest for an uneven distribution of demands over products. The effect of the demand probability factor on the LSP-WSCF—a positive demand in every period leads to longer run-times—

Table 4.22. Factorial design results for large test problems, CLSPL

	CLSPL	
	Ratio	**Time**
DemVar		
low	1.26	1:00:02
high	1.26	1:00:02
InvCosts		
low	1.25	1:00:02
high	1.27	1:00:02
CapUtil		
low	1.28	1:00:02
high	1.22	1:00:02
SetupTime		
low	1.23	1:00:02
high	1.31	1:00:02
DemProb		
every	2.37**	1:00:02
every 2nd	1.21	1:00:02
DemPat		
equal	1.42**	1:00:02
diff sizes	1.28	1:00:02
pos trend	1.11	1:00:02

has also been observed for the small and medium-sized problems, but has only been significant for the small test problems. For the large test problems, the LSP-NSCF behaves in an opposite way: Its run-times are shorter for problems with a positive demand in every period and longer for problems with a positive demand in every second period, but this effect is not statistically significant.

Table 4.23 lists the number of best solutions for each procedure. It also shows that no instance could be solved to proven optimality by the CLSPL. The number of best solutions for all three heuristics are more or less evenly distributed over the different classes. The LSP-NSCF finds more best solutions for low capacity utilization problems, even though its average cost ratio has been better for high capacity utilization problems. Table 4.24 reports on the number of instances that could not be solved by the procedures. Only three instances could not be solved by the LSP-WSCF. These instances could also not be solved by the LSP-NSCF, because the respective figures for the LSP-BOTH remain the same. The LSP-NSCF seems to have some moderate difficulties with problems involving high inventory costs, high capacity

Table 4.23. Number of best solutions and optimally solved instances for large test problems (288 instances)

	LSP WSCF #Best	LSP NSCF #Best	LSP BOTH #Best	CLSPL #Best	CLSPL #Opt
DemVar					
low	86	49	134	8	0
high	99	36	135	8	0
InvCosts					
low	88	45	133	9	0
high	97	40	136	7	0
CapUtil					
low	88	45	132	12	0
high	97	40	137	4	0
SetupTime					
low	87	45	132	12	0
high	98	40	137	4	0
DemProb					
every	92	50	142	0	0
every 2nd	93	35	127	16	0
DemPat					
equal	60	34	93	0	0
diff sizes	65	27	92	4	0
pos trend	60	24	84	12	0
Total	185	85	269	16	0

utilization, high setup times and a positive demand figure in every period. Interestingly, all of these problems have had better or at least equal average cost ratios. However, most striking is the fact that of the 144 problems with a positive demand figure in every period, only four could be solved by the CLSPL. Furthermore, the CLSPL had severe difficulties with highly capacitated problems and high setup times. To sum up, the findings of Table 4.23 and 4.24 reconfirm that the LSP-WSCF clearly outperforms the LSP-NSCF and the CLSPL.

Recapitulating the large test problems, the results indicate that the heuristics are much faster for low capacity utilization and low setup times. No substantial effect on solution quality (cost ratio) of any of the factors could be measured for the two best heuristics, LSP-WSCF and LSP-BOTH. This is

Table 4.24. Number of instances not solved for large test problems (288 instances)

	LSP WSCF #Inst	LSP NSCF #Inst	LSP BOTH #Inst	CLSPL #Inst
DemVar				
low	2	13	2	96
high	1	11	1	95
InvCosts				
low	2	19	2	98
high	1	5	1	93
CapUtil				
low	0	0	0	75
high	3	24	3	116
SetupTime				
low	0	2	0	81
high	3	22	3	110
DemProb				
every	2	16	2	140
every 2nd	1	8	1	51
DemPat				
equal	3	8	3	64
diff sizes	0	7	0	69
pos trend	0	9	0	58
Total	3	24	3	191

due to the fact that these heuristics have been able to find the best solutions for most of the problems. With regard to the long computation time of LSP-BOTH and its only small cost ratio advantage, LSP-WSCF is clearly the favorable heuristic.

4.5.2 Effect of Capacity Adjustment and Time Limit

The heuristics use a time limit when solving model HMO and the scheduling model in step 3 of the scheduling sub-routine. These models are solved using a standard solver (CPLEX). Whenever the standard solver does not terminate within a certain amount of time, the heuristics continue their operation with the best solution found so far. In this section, we investigate the effect of different time limits. We also provide a detailed analysis of the capacity adjustment routine. Two test-beds are inspected, one with large and one with small problem instances.

The large problems consist of 20 products, 6 periods and 6 machines. Ten instances have been created in a similar way to Sect. 4.5.1.1: Demand figures have been generated using a $U(90; 110)$ distribution with the 'different sizes' demand pattern. The demand probability was set to 100%, implying a positive demand figure in every period for all products. Process times have been generated with a $U(5; 15)$ distribution. Initial inventory and back-order volumes have been set to zero, and the initial setup state has been drawn at random. The differences to Sect. 4.5.1.1 are as follows: Both inventory costs and back-order costs are drawn from separate $U(5; 15)$ distributions. Setup times and costs have been created using separate $U(45; 55)$ distributions. Hence, the average setup time is 5% of the per period process time of an average product. Setup times and costs are lower than the relatively high values used in Sect. 4.5.1.1. The capacity has been adjusted to yield a final utilization including setup times of about 90%.

For each of the six variants LSP-WSCF, LSP-WSCF-CA, LSP-NSCF, LSP-NSCF-CA, LSP-BOTH and LSP-BOTH-CA, six different time limits have been investigated: 5 seconds, 15 seconds, 30 seconds, 60 seconds, 120 seconds and 240 seconds. Note that these time limits do not restrict the final computation time of a heuristic, but the time the heuristic spends in each call of the standard solver.

The results, together with the ones of a direct CLSPL implementation with a 10 minute time limit, are presented in Table 4.25. Its columns are analogous to the ones of the previous sections. The ratio compares the result of a procedure with the best known solution of the specific problem over all procedures and time limits. For one instance, the computation time of 'LSP-NSCF-CA 30 s' was about 272 hours. We treat this instance as an outlier and do not consider it for this procedure. For 'LSP-BOTH-CA 30 s', we disregard the instance when reporting computation times but use the solution of 'LSP-WSCF-CA 30 s' for cost ratio calculations.

The standard solver is sometimes not able to find a feasible solution to a sub-problem within a 5 second time limit. In these cases, it is called again with a doubled time limit until it finds a feasible solution. To some extent, this explains why the average computation time of 'LSP-WSCF 15 s' is only two seconds longer than the one of 'LSP-WSCF 5 s'. However, it cannot explain that 'LSP-WSCF 30 s', 'LSP-WSCF 60 s' and 'LSP-WSCF 120 s' are all faster than the two former procedures. This effect can be explained in the following way: With a longer time limit, the standard solver has more time to find good solutions for the scheduling model. A good solution means that the plan can be adhered to as prescribed by the solution of the lot-sizing and scheduling

Table 4.25. Effect of capacity adjustment and time limit on large test problems (10 instances)

Procedure	#Inst	Min	Max	StdDev	Ratio	#Best	Time
LSP-WSCF 5 s	8	6.79	10.09	1.41	8.54	0	00:16:46
LSP-WSCF 15 s	10	3.59	9.99	1.98	7.33	0	00:16:48
LSP-WSCF 30 s	10	1.97	7.41	1.70	4.79	0	00:07:30
LSP-WSCF 60 s	10	1.22	4.01	0.98	1.89	0	00:11:29
LSP-WSCF 120 s	10	1.03	6.52	1.69	1.71	0	00:14:56
LSP-WSCF 240 s	10	1.05	4.01	0.97	1.67	0	00:34:22
LSP-WSCF-CA 5 s	9	6.00	9.99	1.50	8.12	0	00:43:22
LSP-WSCF-CA 15 s	10	1.82	6.95	1.63	4.05	0	01:05:08
LSP-WSCF-CA 30 s	10	1.29	4.96	1.09	2.45	0	01:48:31
LSP-WSCF-CA 60 s	10	1.07	1.91	0.29	1.33	0	02:15:54
LSP-WSCF-CA 120 s	10	1.03	1.73	0.20	1.17	0	02:27:48
LSP-WSCF-CA 240 s	10	1.00	1.41	0.12	1.12	1	07:00:51
LSP-NSCF 5 s	4	7.33	9.72	1.08	8.74	0	00:07:46
LSP-NSCF 15 s	9	1.60	7.59	2.25	4.60	0	00:48:18
LSP-NSCF 30 s	10	1.16	13.50	3.68	3.68	0	00:19:39
LSP-NSCF 60 s	10	1.16	5.14	1.21	2.06	0	00:11:42
LSP-NSCF 120 s	10	1.03	1.45	0.12	1.20	0	00:14:23
LSP-NSCF 240 s	10	1.00	1.24	0.09	1.10	0	00:22:55
LSP-NSCF-CA 5 s	10	5.45	9.72	1.58	7.85	0	00:56:29
LSP-NSCF-CA 15 s	10	1.60	6.94	1.65	3.54	0	03:27:16
LSP-NSCF-CA 30 s	9	1.16	5.36	1.34	2.35	0	03:51:42
LSP-NSCF-CA 60 s	10	1.11	1.79	0.22	1.30	0	05:50:23
LSP-NSCF-CA 120 s	10	1.00	1.16	0.05	1.05	1	14:11:17
LSP-NSCF-CA 240 s	10	1.00	1.05	0.02	1.01	8	27:35:28
LSP-BOTH 5 s	8	6.79	9.99	1.31	8.20	0	00:28:57
LSP-BOTH 15 s	10	1.60	9.16	2.56	5.05	0	01:01:23
LSP-BOTH 30 s	10	1.16	4.12	1.07	2.53	0	00:27:09
LSP-BOTH 60 s	10	1.16	2.70	0.53	1.58	0	00:23:11
LSP-BOTH 120 s	10	1.03	1.19	0.06	1.11	0	00:29:19
LSP-BOTH 240 s	10	1.00	1.24	0.08	1.09	0	00:57:17
LSP-BOTH-CA 5 s	10	5.45	9.45	1.22	7.14	0	01:35:31
LSP-BOTH-CA 15 s	10	1.60	4.61	1.16	2.87	0	04:32:24
LSP-BOTH-CA 30 s	10	1.16	3.09	0.69	2.00	0	05:45:59
LSP-BOTH-CA 60 s	10	1.07	1.79	0.21	1.22	0	08:06:17
LSP-BOTH-CA 120 s	10	1.00	1.13	0.04	1.04	1	16:39:05
LSP-BOTH-CA 240 s	10	1.00	1.05	0.02	1.01	9	34:36:19
CLSPL	6	4.74	10.00	1.79	6.62	0	00:10:01

Table 4.26. Average number of backtrackings for LSP-WSCF and LSP-NSCF with different time limits

Time limit	LSP-WSCF	LSP-NSCF
5 s	79.0	61.8
15 s	31.2	97.8
30 s	3.9	19.1
60 s	3.0	2.3
120 s	0.8	0.8
240 s	2.4	0.1

model and no additional back-orders have to be scheduled. A shorter time limit might lead to worse solutions that result in higher back-orders. Higher back-orders increase the probability that no feasible plan can be found in the last period, and a period backtracking has to be performed to reach feasibility. Ultimately, this might result in longer total run-times of the procedure. Table 4.26 shows the average number of backtrackings for the LSP-WSCF with different time limits, supporting our reasoning. A similar picture is given by the figures for LSP-NSCF. For LSP-BOTH, both values would have to be added.

Figure 4.16 illustrates the effect of different time limits on the cost ratio. The left side shows the cost ratios of procedures without capacity adjustment (namely LSP-WSCF, LSP-NSCF and LSP-BOTH), the right side the ratios

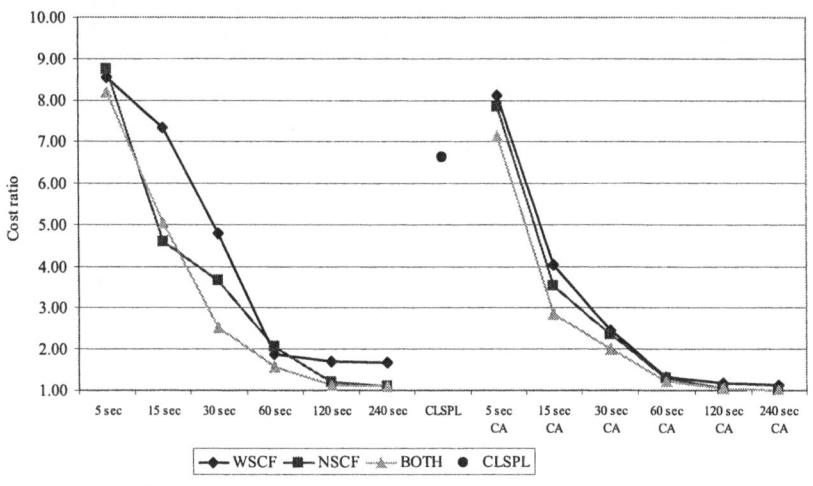

Fig. 4.16. Effect of time limit on cost ratio for large test problems

of procedures with capacity adjustment (LSP-WSCF-CA, LSP-NSCF-CA and LSP-BOTH-CA). As a comparison, the cost ratio of the 10 minute time limited CLSPL is depicted in the middle.

The NSCF procedures lead to better results than the WSCF ones. This is in contrast to the earlier findings. Moreover, for a 15 second time limit, 'LSP-NSCF 15 s' leads to a better average cost ratio than 'LSP-BOTH 15 s'. This is because 'LSP-NSCF 15 s' solves only nine instances. The tenth instance is solved by 'LSP-WSCF 15 s' with a cost ratio of 9.16, which also raises the average ratio for 'LSP-BOTH 15 s'.

We can state that the longer the time limit, the better the solution quality. However, the cost ratio improvement of a time limit doubling decelerates the higher the time limit. A prolongation over 120 seconds appears not to lead to substantially better solutions.

In general, the longer the time limit, the longer the total computation time of the procedures. The computation times of procedures without capacity adjustment are illustrated in Fig. 4.17, of procedures with capacity adjustment in Fig. 4.18 (notice the different time-scales).

With the above-mentioned exception for the non-CA procedures at a short time limit, a time limit extension entails substantially longer computation times. Given the relatively small improvement of the cost ratio, a time limit of more than 120 seconds does not seem appropriate. In any case, the average computation time of 7:00:51 for 'LSP-WSCF-CA 240 s' is caused by a single instance running for more than 35 hours. Without this instance, the average run-time is 3:47:53.

Figures 4.19 and 4.20 portray the same results from a different point of view: The specific CA procedures and their non-CA counterparts are displayed next to each other, allowing a detailed analysis of the capacity adjustment routine.

The cost ratio improves substantially when the capacity adjustment routine is invoked. However, the computation time increases as well. The longer the time limit, the stronger is this effect. For example, with a time limit of 240 seconds, LSP-BOTH has an average computation time of about one hour. Incorporating capacity adjustment, LSP-BOTH-CA has an average computation time of more than 34 hours. However, the increase of computation time is less steep for LSP-WSCF-CA. This can be explained by the fact that the main objective of the scheduling sub-routine of LSP-WSCF-CA is to schedule the complete production volume as given by the solutions of the lot-sizing and scheduling model. In contrast, the scheduling sub-routine of LSP-NSCF-CA can lead to leftover capacity even though the recommended production volumes are not fulfilled (see Sect. 4.4.2.5). Since a capacity adjustment is invoked whenever the recommended production volume is not produced completely,

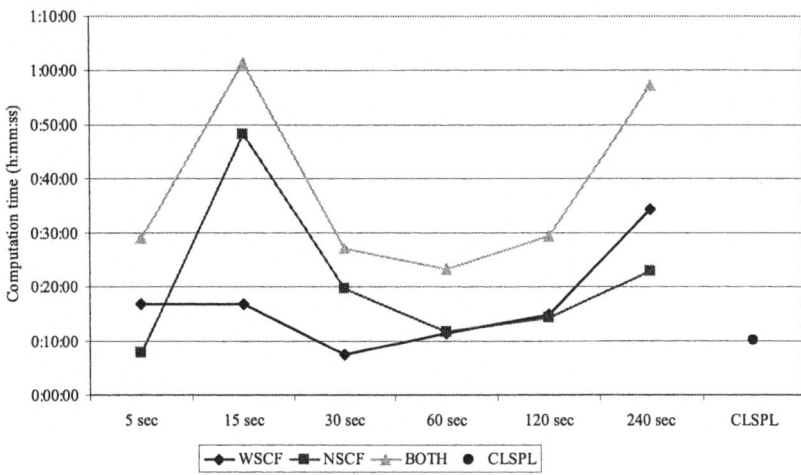

Fig. 4.17. Effect of time limit on computation time for large test problems, non-CA procedures

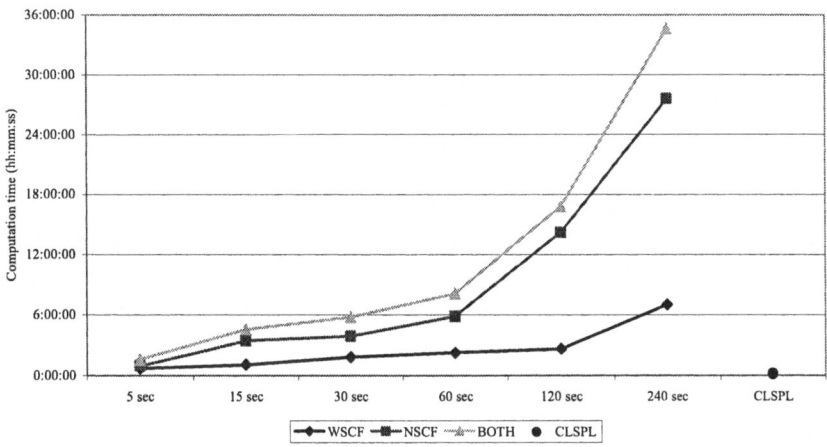

Fig. 4.18. Effect of time limit on computation time for large test problems, CA procedures

it is obvious that the LSP-WSCF-CA will perform fewer runs on average and lead to shorter computation times.

The small test problems consist of 7 products, 4 periods and 4 machines. Ten instances have been created as described above. Four of them could be solved to proven optimality using the 10 minute time limited CLSPL. To allow

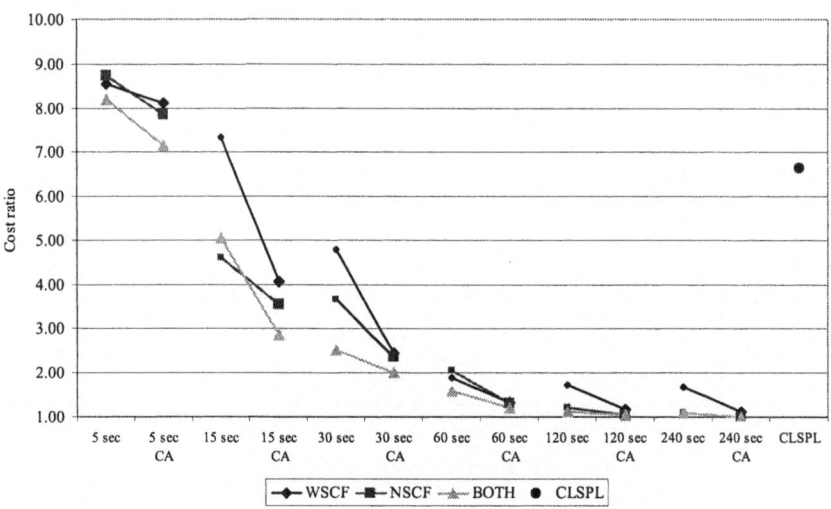

Fig. 4.19. Effect of capacity adjustment on cost ratio for large test problems

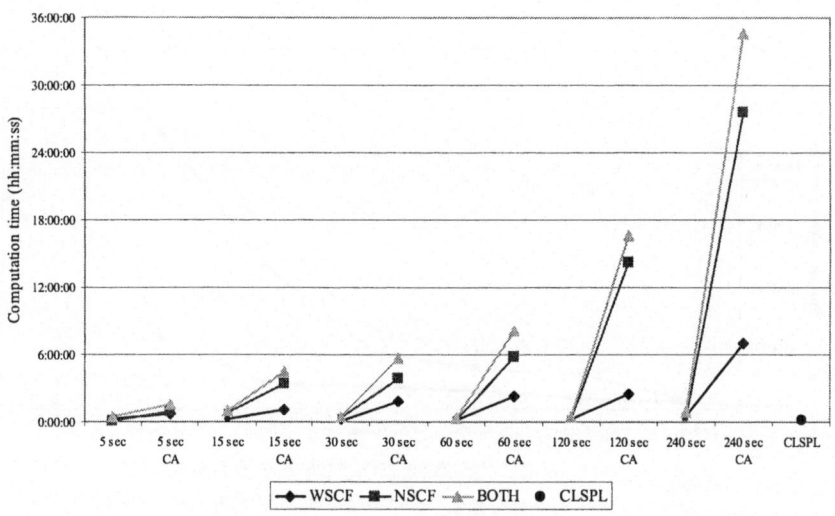

Fig. 4.20. Effect of capacity adjustment on computation time for large test problems

a comparison with optimal solutions, we report on these instances only. Table 4.27 shows the results for all procedures and time limits.

No matter which time limit is used, all heuristics find the optimal solution to three of the four problems: The small deviance of the average cost ratio is caused by a single instance only. Also the difference in computation times is

Table 4.27. Effect of capacity adjustment and time limit on small test problems (4 instances)

Procedure	#Inst	Min	Max	StdDev	Ratio	#Best	Time
LSP-WSCF 5 s	4	1.00	1.16	0.08	1.04	3	0:00:04
LSP-WSCF 15 s	4	1.00	1.16	0.08	1.04	3	0:00:04
LSP-WSCF 30 s	4	1.00	1.16	0.08	1.04	3	0:00:04
LSP-WSCF 60 s	4	1.00	1.16	0.08	1.04	3	0:00:04
LSP-WSCF 120 s	4	1.00	1.16	0.08	1.04	3	0:00:04
LSP-WSCF 240 s	4	1.00	1.16	0.08	1.04	3	0:00:04
LSP-WSCF-CA 5 s	4	1.00	1.16	0.08	1.04	3	0:00:04
LSP-WSCF-CA 15 s	4	1.00	1.16	0.08	1.04	3	0:00:04
LSP-WSCF-CA 30 s	4	1.00	1.16	0.08	1.04	3	0:00:04
LSP-WSCF-CA 60 s	4	1.00	1.16	0.08	1.04	3	0:00:04
LSP-WSCF-CA 120 s	4	1.00	1.16	0.08	1.04	3	0:00:04
LSP-WSCF-CA 240 s	4	1.00	1.16	0.08	1.04	3	0:00:04
LSP-NSCF 5 s	4	1.00	1.08	0.04	1.02	3	0:00:03
LSP-NSCF 15 s	4	1.00	1.08	0.04	1.02	3	0:00:03
LSP-NSCF 30 s	4	1.00	1.08	0.04	1.02	3	0:00:03
LSP-NSCF 60 s	4	1.00	1.08	0.04	1.02	3	0:00:04
LSP-NSCF 120 s	4	1.00	1.08	0.04	1.02	3	0:00:03
LSP-NSCF 240 s	4	1.00	1.08	0.04	1.02	3	0:00:03
LSP-NSCF-CA 5 s	4	1.00	1.08	0.04	1.02	3	0:00:13
LSP-NSCF-CA 15 s	4	1.00	1.08	0.04	1.02	3	0:00:14
LSP-NSCF-CA 30 s	4	1.00	1.08	0.04	1.02	3	0:00:14
LSP-NSCF-CA 60 s	4	1.00	1.08	0.04	1.02	3	0:00:14
LSP-NSCF-CA 120 s	4	1.00	1.08	0.04	1.02	3	0:00:13
LSP-NSCF-CA 240 s	4	1.00	1.08	0.04	1.02	3	0:00:13
LSP-BOTH 5 s	4	1.00	1.08	0.04	1.02	3	0:00:07
LSP-BOTH 15 s	4	1.00	1.08	0.04	1.02	3	0:00:07
LSP-BOTH 30 s	4	1.00	1.08	0.04	1.02	3	0:00:07
LSP-BOTH 60 s	4	1.00	1.08	0.04	1.02	3	0:00:07
LSP-BOTH 120 s	4	1.00	1.08	0.04	1.02	3	0:00:07
LSP-BOTH 240 s	4	1.00	1.08	0.04	1.02	3	0:00:07
LSP-BOTH-CA 5 s	4	1.00	1.08	0.04	1.02	3	0:00:17
LSP-BOTH-CA 15 s	4	1.00	1.08	0.04	1.02	3	0:00:18
LSP-BOTH-CA 30 s	4	1.00	1.08	0.04	1.02	3	0:00:17
LSP-BOTH-CA 60 s	4	1.00	1.08	0.04	1.02	3	0:00:18
LSP-BOTH-CA 120 s	4	1.00	1.08	0.04	1.02	3	0:00:17
LSP-BOTH-CA 240 s	4	1.00	1.08	0.04	1.02	3	0:00:17
CLSPL	4	1.00	1.00	0.00	1.00	4	0:01:57

marginal. This is due to the fact that the time limit has never been reached for any of the heuristics. Thus, for small test problems, a longer time limit does not lead to better solutions but also does not impair the heuristics as their computation times are not increased. For the four instances tested here, the capacity adjustment routine also has no impact on cost ratio and only a small impact on run-time. The LSP-WSCF-CA variants even terminate after a single run, leading to the same solution and run-time as their LSP-WSCF counterparts. For a more detailed picture of the capacity adjustment routine on small test instances, we refer to Sect. 4.5.1.2.

4.5.3 Effect of Problem Size

In this section, we analyze the effect of different problem sizes on the heuristics. We systematically vary the number of products, the number of periods and the number of machines covered by the test problems.

To study the effect of varying the **number of products**, eight scenarios with 2, 3, 4, 5, 6, 8, 10 and 15 products are investigated. Each scenario consists of 10 instances that have been generated using the same parameters as described in Sect. 4.5.2. As a comparison, we employ two versions of the direct CLSPL implementation. One is time limited by 10 minutes, the other one by 300 minutes.

The results are depicted in Tables 4.28 and 4.29. The columns are analogous to the ones of the previous sections. An additional column has been added for the two CLSPL versions that shows the number of instances solved to proven optimality. The ratio compares the result of a procedure with the best known solution to the specific problem over all procedures.

The average cost ratios are displayed in Fig. 4.21. The cost ratios for the non-CA procedures LSP-WSCF, LSP-NSCF and LSP-BOTH are shown on the left side, the ones for the CA procedures LSP-WSCF-CA, LSP-NSCF-CA and LSP-BOTH-CA on the right side. The results for the two CLSPL variants are underlaid on both sides.

Leaving out the very small problems with only two products, there is a general trend for all heuristics to obtain better solutions the more products are considered. This can be explained by the deteriorating performance of the direct CLSPL implementations. The more products are considered, the worse the comparative cost ratio and thus the better the heuristics. While for small test problems, the 300 minute time limited CLSPL can solve a majority of instances to proven optimality, it cannot find good solutions for larger problems.

Another insight is that the WSCF procedures seem to perform better when only a small number of products is considered, while for a higher number of products, the NSCF solutions are better: As can be seen in Fig. 4.21, for the

Table 4.28. Effect of varying the number of products for non-CA procedures and CLSPL 10 min (10 instances each)

#Products	#Inst	Min	Max	StdDev	Ratio	#Best	Time	#Opt
LSP-WSCF								
2	10	1.00	1.23	0.09	1.06	6	0:00:00	
3	10	1.00	1.31	0.11	1.16	2	0:00:01	
4	10	1.00	1.43	0.12	1.15	1	0:00:09	
5	10	1.00	1.24	0.07	1.07	2	0:00:05	
6	10	1.00	1.21	0.07	1.13	0	0:00:09	
8	10	1.00	1.27	0.10	1.07	4	0:00:33	
10	10	1.00	1.19	0.07	1.09	1	0:03:18	
15	10	1.02	1.28	0.08	1.11	0	0:10:42	
LSP-NSCF								
2	10	1.00	1.52	0.19	1.12	6	0:00:00	
3	10	1.00	1.31	0.11	1.18	2	0:00:01	
4	10	1.00	1.43	0.14	1.19	1	0:00:07	
5	10	1.00	1.24	0.08	1.08	2	0:00:04	
6	10	1.00	1.18	0.06	1.11	0	0:00:07	
8	10	1.00	1.08	0.03	1.03	5	0:00:27	
10	10	1.00	1.14	0.05	1.04	3	0:01:10	
15	10	1.00	1.16	0.05	1.05	2	0:10:53	
LSP-BOTH								
2	10	1.00	1.23	0.09	1.06	6	0:00:00	
3	10	1.00	1.31	0.11	1.16	2	0:00:02	
4	10	1.00	1.43	0.12	1.15	1	0:00:16	
5	10	1.00	1.24	0.07	1.07	2	0:00:09	
6	10	1.00	1.18	0.06	1.11	0	0:00:16	
8	10	1.00	1.08	0.03	1.03	5	0:01:00	
10	10	1.00	1.14	0.05	1.04	3	0:04:28	
15	10	1.00	1.16	0.05	1.05	2	0:21:36	
CLSPL 10 min								
2	10	1.00	1.00	0.00	1.00	10	0:00:02	10
3	10	1.00	1.00	0.00	1.00	10	0:04:12	7
4	10	1.00	1.04	0.01	1.00	8	0:07:11	3
5	10	1.00	1.14	0.05	1.02	6	0:07:35	3
6	10	1.00	1.26	0.09	1.10	2	0:10:00	0
8	10	1.02	1.41	0.12	1.24	0	0:10:00	0
10	10	1.09	1.59	0.16	1.36	0	0:10:01	0
15	10	1.28	3.81	0.81	2.27	0	0:10:01	0

Table 4.29. Effect of varying the number of products for CA procedures and CLSPL 300 min (10 instances each)

#Products	#Inst	Min	Max	StdDev	Ratio	#Best	Time	#Opt
LSP-WSCF-CA								
2	10	1.00	1.23	0.09	1.06	6	0:00:00	
3	10	1.00	1.31	0.11	1.16	2	0:00:01	
4	10	1.00	1.43	0.12	1.15	1	0:00:11	
5	10	1.00	1.24	0.07	1.07	2	0:00:05	
6	10	1.00	1.20	0.06	1.09	0	0:00:31	
8	10	1.00	1.14	0.05	1.04	4	0:02:50	
10	10	1.00	1.13	0.04	1.03	3	1:17:50	
15	10	1.00	1.12	0.04	1.05	2	1:03:32	
LSP-NSCF-CA								
2	10	1.00	1.52	0.19	1.12	6	0:00:01	
3	10	1.00	1.31	0.11	1.18	2	0:00:02	
4	10	1.00	1.43	0.12	1.17	1	0:00:10	
5	10	1.00	1.24	0.08	1.08	2	0:00:06	
6	10	1.00	1.18	0.06	1.06	2	0:00:36	
8	10	1.00	1.00	0.00	1.00	9	0:12:51	
10	10	1.00	1.02	0.01	1.00	8	2:01:06	
15	10	1.00	1.01	0.00	1.00	8	8:33:22	
LSP-BOTH-CA								
2	10	1.00	1.23	0.09	1.06	6	0:00:02	
3	10	1.00	1.31	0.11	1.16	2	0:00:03	
4	10	1.00	1.43	0.12	1.15	1	0:00:21	
5	10	1.00	1.24	0.07	1.07	2	0:00:11	
6	10	1.00	1.18	0.06	1.06	2	0:01:07	
8	10	1.00	1.00	0.00	1.00	9	0:15:41	
10	10	1.00	1.02	0.01	1.00	9	3:18:56	
15	10	1.00	1.00	0.00	1.00	10	9:36:53	
CLSPL 300 min								
2	10	1.00	1.00	0.00	1.00	10	0:00:02	10
3	10	1.00	1.00	0.00	1.00	10	0:45:09	9
4	10	1.00	1.00	0.00	1.00	10	2:09:52	6
5	10	1.00	1.00	0.00	1.00	10	3:02:29	4
6	10	1.00	1.05	0.02	1.01	8	3:53:23	3
8	10	1.00	1.23	0.07	1.06	1	5:00:00	0
10	10	1.00	1.23	0.08	1.13	1	5:00:00	0
15	10	1.14	1.25	0.03	1.17	0	5:00:01	0

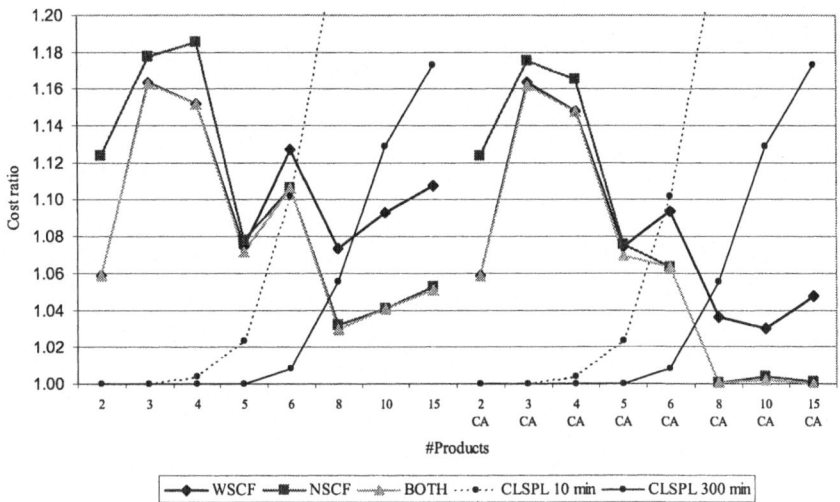

Fig. 4.21. Effect of varying the number of products on cost ratio

small problems, the BOTH solutions are taken from its WSCF component, whereas for the larger problems, they are taken from its NSCF component. This is true for the procedures with and without capacity adjustment.

The average computation times of procedures without capacity adjustment, along with the ones of the 10 minute time limited CLSPL, are shown in Fig. 4.22. Figure 4.23 shows the average computation times of procedures with capacity adjustment along with the ones of the 300 minute time limited CLSPL.

Though at different time scales, both figures show similar curves. The more products are considered, the longer the computation time. The increase in computation time appears to be exponential. For the CLSPL procedures, the picture is somewhat misleading. The linear curve stems from the fact that with an increasing number of products, fewer problems can be solved to optimality. Hence, the average computation time converges to the time limit. Concerning the heuristics, the routines with capacity adjustment have especially long computation times. As already observed in Sect. 4.5.2, the runtime increase of LSP-WSCF-CA is less steep than the one of LSP-NSCF-CA. LSP-WSCF-CA even shows a shorter average computation time for problems with 15 products than for problems with 10 products. This is because for one problem with 10 products, LSP-WSCF-CA has a run time of 11:44:34. Without this instance, the average run-time decreases from 1:17:50 as listed in Table 4.29 to 0:08:12.

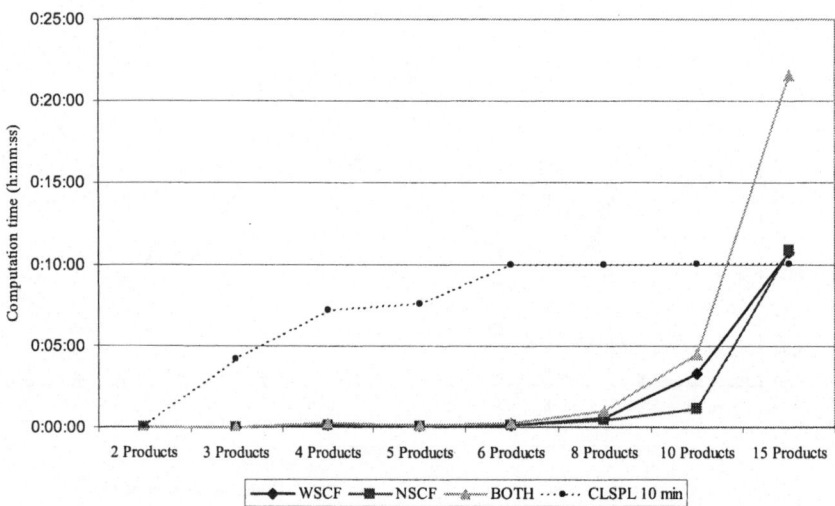

Fig. 4.22. Effect of varying the number of products on computation time, non-CA procedures and CLSPL 10 min

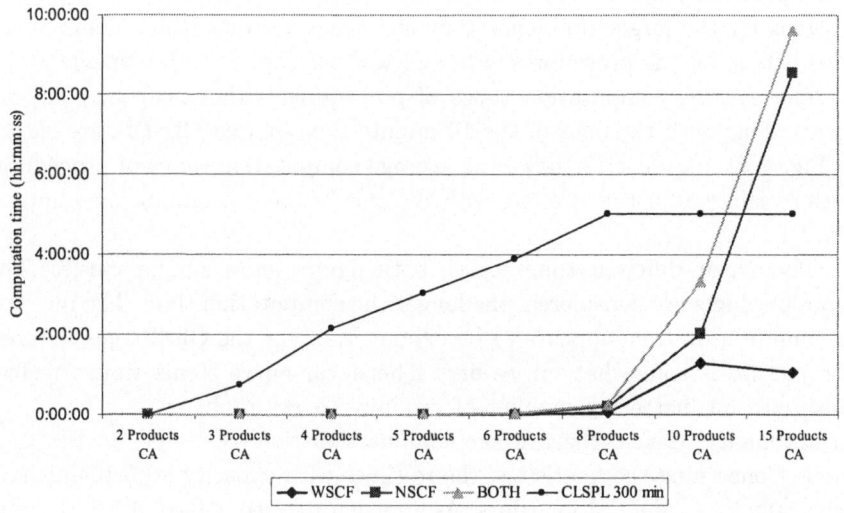

Fig. 4.23. Effect of varying the number of products on computation time, CA procedures and CLSPL 300 min

The effect of varying the **number of periods** is investigated using six scenarios, consisting of 2, 4, 6, 8, 10 and 15 periods. Each scenario consists of 10 instances generated as described in Sect. 4.5.2. A direct implementation of the CLSPL with a time limit of 10 minutes serves as a comparison.

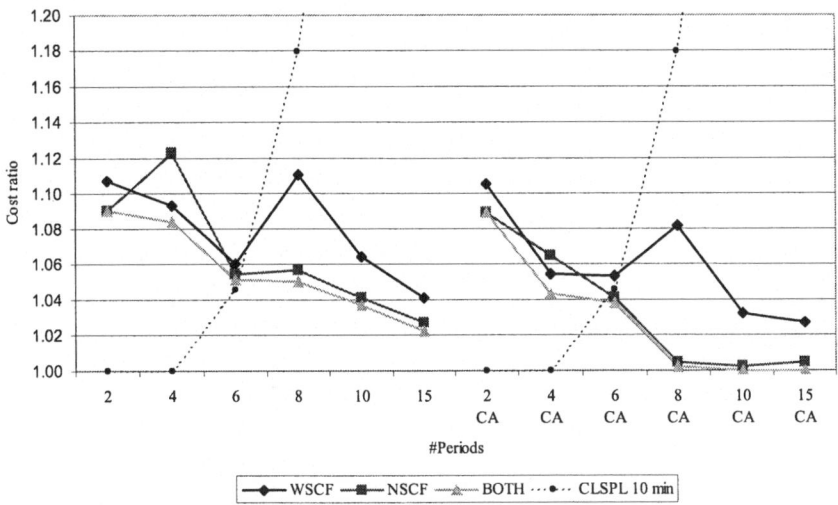

Fig. 4.24. Effect of varying the number of periods on cost ratio

Tables 4.30 and 4.31 show the results. The average cost ratios are displayed in Fig. 4.24.

In general, the more periods are considered, the better the cost ratio of the heuristics. This trend is even more evident here than for the number of products and can again be explained by the deteriorating performance of the CLSPL. The non-CA procedures lead to lower average cost ratios for problems with 15 periods when compared with problems covering 15 products—however, the difference is only statistically significant for LSP-WSCF (at an error level of 5% using an unpaired t-test).

The average computation times are shown in Fig. 4.25 for procedures without capacity adjustment and in Fig. 4.26 for procedures with capacity adjustment. Both times, the average computation times for the 10 minute time limited CLSPL are underlaid.

The computation times for the procedures without capacity adjustment are similar to the ones for varying the number of products: A change in the number of periods has a similar effect on the computation time of these heuristics as a change in the number of products. However, for the heuristics with capacity adjustment, the computation time curve for an increasing number of periods is less steep than the one for an increasing number of products. An increase in periods also appears to lead to exponentially increasing runtimes, but the factor is lower. Thus, in terms of run-time, it seems that the CA heuristics can more easily cope with an increase in the number of periods than with an increase in the number of products. Similar to the above results,

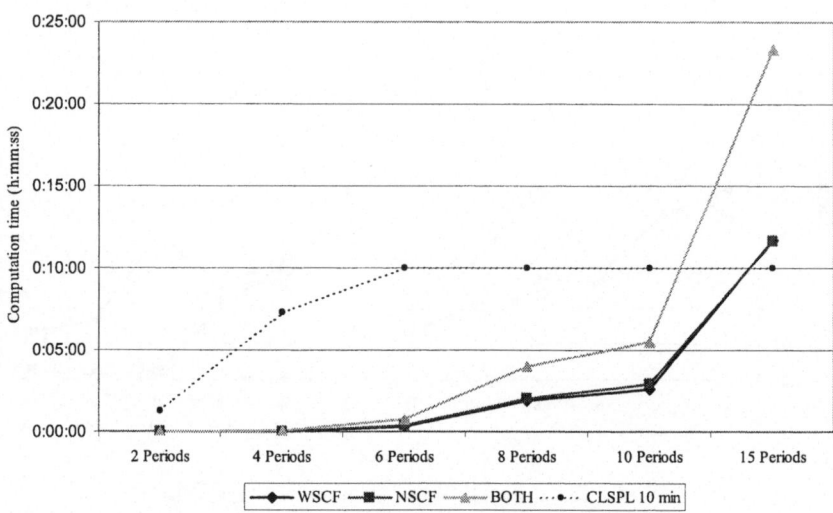

Fig. 4.25. Effect of varying the number of periods on computation time, non-CA procedures and CLSPL

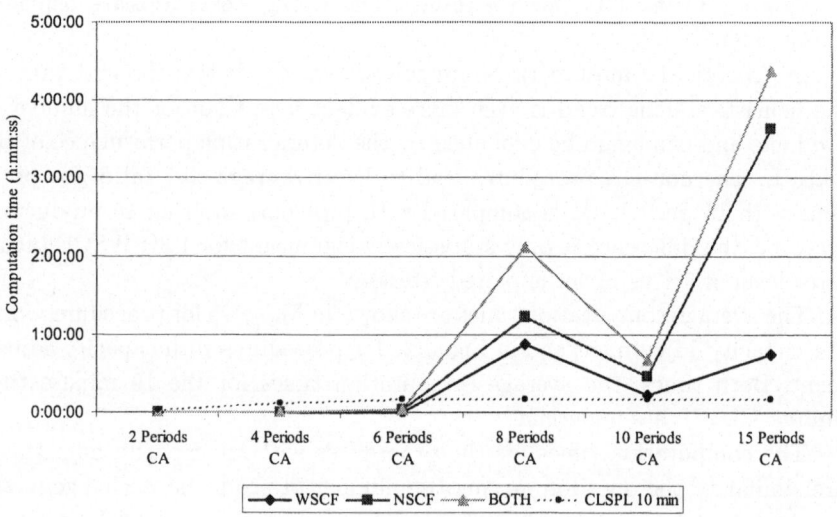

Fig. 4.26. Effect of varying the number of periods on computation time, CA procedures and CLSPL

Table 4.30. Effect of varying the number of periods (10 instances each)

#Periods	#Ins	Min	Max	StdDev	Ratio	#Best	Time
LSP-WSCF							
2	10	1.00	1.41	0.13	1.11	4	0:00:00
4	10	1.00	1.28	0.10	1.09	3	0:00:02
6	10	1.00	1.16	0.06	1.06	3	0:00:20
8	10	1.00	1.36	0.14	1.11	4	0:01:54
10	10	1.00	1.15	0.06	1.06	4	0:02:35
15	10	1.00	1.15	0.05	1.04	3	0:11:41
LSP-NSCF							
2	10	1.00	1.41	0.14	1.09	5	0:00:00
4	10	1.00	1.37	0.12	1.12	2	0:00:02
6	10	1.00	1.16	0.06	1.05	3	0:00:24
8	10	1.00	1.26	0.08	1.06	3	0:02:03
10	10	1.00	1.13	0.05	1.04	4	0:02:54
15	10	1.00	1.09	0.03	1.03	4	0:11:39
LSP-BOTH							
2	10	1.00	1.41	0.14	1.09	5	0:00:01
4	10	1.00	1.26	0.09	1.08	3	0:00:03
6	10	1.00	1.16	0.06	1.05	5	0:00:44
8	10	1.00	1.22	0.08	1.05	5	0:03:57
10	10	1.00	1.13	0.05	1.04	5	0:05:29
15	10	1.00	1.09	0.04	1.02	6	0:23:20
LSP-WSCF-CA							
2	10	1.00	1.40	0.13	1.11	4	0:00:00
4	10	1.00	1.11	0.05	1.05	3	0:00:03
6	10	1.00	1.15	0.06	1.05	3	0:00:23
8	10	1.00	1.32	0.11	1.08	4	0:53:10
10	10	1.00	1.14	0.05	1.03	5	0:13:16
15	10	1.00	1.08	0.03	1.03	4	0:44:52
LSP-NSCF-CA							
2	10	1.00	1.40	0.13	1.09	5	0:00:01
4	10	1.00	1.16	0.06	1.06	2	0:00:08
6	10	1.00	1.13	0.04	1.04	3	0:02:08
8	10	1.00	1.02	0.01	1.00	7	1:13:44
10	10	1.00	1.03	0.01	1.00	9	0:27:49
15	10	1.00	1.03	0.01	1.00	7	3:37:24

Table 4.31. Effect of varying the number of periods (10 instances each, *continued*)

#Periods	#Ins	Min	Max	StdDev	Ratio	#Best	Time	#Opt
LSP-BOTH-CA								
2	10	1.00	1.40	0.13	1.09	5	0:00:02	
4	10	1.00	1.10	0.04	1.04	3	0:00:11	
6	10	1.00	1.13	0.05	1.04	5	0:02:31	
8	10	1.00	1.02	0.01	1.00	9	2:06:54	
10	10	1.00	1.00	0.00	1.00	10	0:41:06	
15	10	1.00	1.00	0.00	1.00	10	4:22:16	
CLSPL 10 min								
2	10	1.00	1.00	0.00	1.00	10	0:01:17	9
4	10	1.00	1.00	0.00	1.00	10	0:07:15	5
6	10	1.00	1.33	0.11	1.05	7	0:10:00	0
8	10	1.00	1.41	0.16	1.18	1	0:10:00	0
10	10	1.24	1.72	0.15	1.41	0	0:10:00	0
15	10	1.11	2.65	0.49	1.76	0	0:10:00	0

the computation time curve for LSP-WSCF-CA is less steep than the one for the other heuristics with capacity adjustment.

Another study is conducted to investigate the influence of the **number of machines** on the procedures. Ten scenarios with 2, 4, 6, 8, 10, 15, 20, 30, 50 and 100 machines have been generated. Each scenario consists of 10 instances, created as described in Sect. 4.5.2. As above, a direct implementation of the CLSPL with a time limit of 10 minutes is used as a comparison.

The results are presented in Tables 4.32, 4.33 and 4.34. Figure 4.27 shows the average cost ratios.

The results for the cost ratio seem inconsistent. It appears that the heuristics find better solutions the more machines are considered. Their average cost ratios are also low for a medium number of machines (10–15), which is probably due to the relatively poor performance of the CLSPL for these problems. For large problems with many machines, all procedures—the heuristics and the CLSPL—are able to find very good solutions. Obviously, it is easy to schedule the demand of six products on 50 or 100 machines, even though the capacity including setups is about 90%. Nevertheless, optimality of the CLSPL solutions could not be proven for any of the instances and the CLSPL could not solve one of the 10 instances with a 10 minute time limit.

In contrast to the findings for the number of products, it appears that the WSCF heuristics perform better when the number of machines is high, whereas the NSCF heuristics lead to better cost ratios when the number of machines is low.

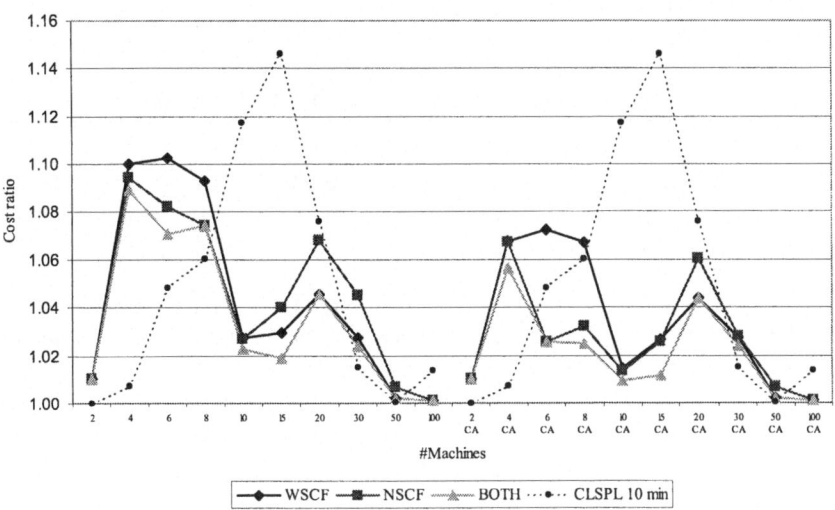

Fig. 4.27. Effect of varying the number of machines on cost ratio

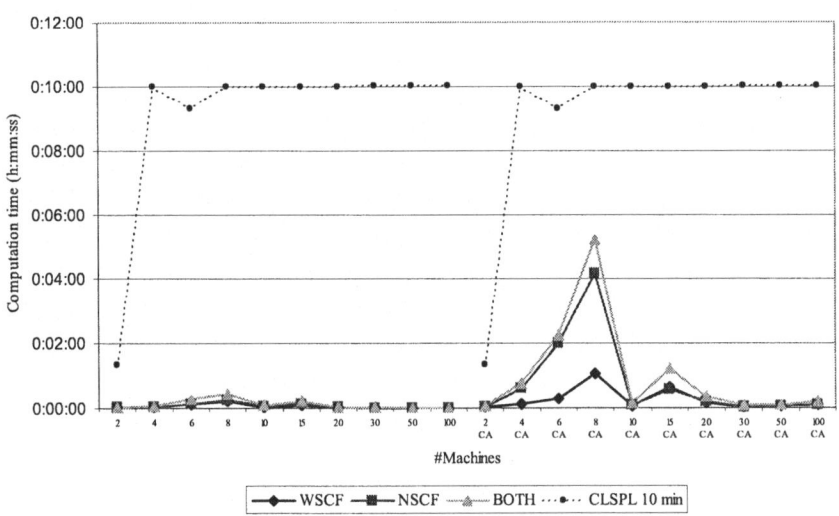

Fig. 4.28. Effect of varying the number of machines on computation time

The average computation times are depicted in Fig. 4.28. For all heuristics, the average run-time is low in every scenario. It increases slightly for a medium number of machines, but decreases again for a high number of machines. It appears that the heuristics are generally unaffected by the number of machines considered. As above, the average run-time of LSP-WSCF-CA is

Table 4.32. Effect of varying the number of machines for non-CA procedures (10 instances each)

#Machines	#Inst	Min	Max	StdDev	Ratio	#Best	Time
LSP-WSCF							
2	10	1.00	1.07	0.02	1.01	8	0:00:01
4	10	1.00	1.32	0.11	1.10	3	0:00:02
6	10	1.03	1.28	0.07	1.10	0	0:00:08
8	10	1.00	1.27	0.10	1.09	0	0:00:13
10	10	1.00	1.13	0.04	1.03	5	0:00:03
15	10	1.00	1.08	0.04	1.03	5	0:00:06
20	10	1.00	1.14	0.06	1.05	6	0:00:01
30	10	1.00	1.09	0.03	1.03	3	0:00:00
50	10	1.00	1.02	0.01	1.00	9	0:00:00
100	10	1.00	1.01	0.00	1.00	8	0:00:00
LSP-NSCF							
2	10	1.00	1.07	0.02	1.01	8	0:00:01
4	10	1.00	1.32	0.10	1.09	2	0:00:02
6	10	1.00	1.19	0.06	1.08	1	0:00:09
8	10	1.00	1.25	0.09	1.07	2	0:00:15
10	10	1.00	1.13	0.04	1.03	4	0:00:03
15	10	1.00	1.21	0.07	1.04	5	0:00:07
20	10	1.00	1.14	0.05	1.07	3	0:00:01
30	10	1.00	1.15	0.05	1.05	3	0:00:01
50	10	1.00	1.07	0.02	1.01	9	0:00:00
100	10	1.00	1.01	0.00	1.00	8	0:00:00
LSP-BOTH							
2	10	1.00	1.07	0.02	1.01	8	0:00:03
4	10	1.00	1.32	0.10	1.09	3	0:00:04
6	10	1.00	1.19	0.05	1.07	1	0:00:17
8	10	1.00	1.25	0.09	1.07	2	0:00:27
10	10	1.00	1.13	0.04	1.02	5	0:00:06
15	10	1.00	1.07	0.03	1.02	5	0:00:13
20	10	1.00	1.14	0.06	1.05	6	0:00:02
30	10	1.00	1.09	0.03	1.02	4	0:00:01
50	10	1.00	1.02	0.01	1.00	9	0:00:00
100	10	1.00	1.01	0.00	1.00	8	0:00:01

shorter than the one of LSP-NSCF-CA. The CLSPL uses its full time limit in all larger instances, leading to substantially longer computation times. The heuristics are able to find comparable or better solutions in a substantially shorter amount of time.

Table 4.33. Effect of varying the number of machines for CA procedures (10 instances each)

#Machines	#Inst	Min	Max	StdDev	Ratio	#Best	Time
LSP-WSCF-CA							
2	10	1.00	1.07	0.02	1.01	8	0:00:02
4	10	1.00	1.21	0.08	1.07	3	0:00:09
6	10	1.03	1.13	0.03	1.07	0	0:00:17
8	10	1.00	1.27	0.08	1.07	1	0:01:05
10	10	1.00	1.05	0.02	1.01	6	0:00:04
15	10	1.00	1.08	0.03	1.03	5	0:00:38
20	10	1.00	1.14	0.06	1.04	6	0:00:10
30	10	1.00	1.09	0.03	1.03	3	0:00:02
50	10	1.00	1.02	0.01	1.00	9	0:00:03
100	10	1.00	1.01	0.00	1.00	8	0:00:06
LSP-NSCF-CA							
2	10	1.00	1.07	0.02	1.01	8	0:00:02
4	10	1.00	1.16	0.05	1.07	2	0:00:37
6	10	1.00	1.09	0.03	1.03	5	0:02:00
8	10	1.00	1.10	0.04	1.03	4	0:04:10
10	10	1.00	1.04	0.02	1.01	6	0:00:06
15	10	1.00	1.21	0.07	1.03	8	0:00:35
20	10	1.00	1.14	0.05	1.06	3	0:00:12
30	10	1.00	1.09	0.03	1.03	3	0:00:03
50	10	1.00	1.07	0.02	1.01	9	0:00:03
100	10	1.00	1.01	0.00	1.00	8	0:00:06
LSP-BOTH-CA							
2	10	1.00	1.07	0.02	1.01	8	0:00:03
4	10	1.00	1.16	0.06	1.06	3	0:00:46
6	10	1.00	1.09	0.03	1.03	5	0:02:16
8	10	1.00	1.10	0.04	1.03	5	0:05:15
10	10	1.00	1.04	0.02	1.01	7	0:00:10
15	10	1.00	1.07	0.03	1.01	8	0:01:13
20	10	1.00	1.14	0.06	1.04	6	0:00:21
30	10	1.00	1.09	0.03	1.02	4	0:00:05
50	10	1.00	1.02	0.01	1.00	9	0:00:06
100	10	1.00	1.01	0.00	1.00	8	0:00:12

Table 4.34. Effect of varying the number of machines for CLSPL (10 instances each)

#Machines	#Inst	Min	Max	StdDev	Ratio	#Best	Time	#Opt
CLSPL 10 min								
2	10	1.00	1.00	0.00	1.00	10	0:01:21	10
4	10	1.00	1.06	0.02	1.01	8	0:10:00	0
6	10	1.00	1.26	0.08	1.05	5	0:09:20	1
8	10	1.00	1.23	0.09	1.06	5	0:10:00	0
10	10	1.00	1.54	0.17	1.12	3	0:10:00	0
15	10	1.00	1.34	0.12	1.15	2	0:10:00	0
20	10	1.00	1.31	0.11	1.08	4	0:10:01	0
30	10	1.00	1.09	0.03	1.02	7	0:10:01	0
50	10	1.00	1.00	0.00	1.00	9	0:10:01	0
100	9	1.00	1.11	0.04	1.01	7	0:10:02	0

4.6 Summary

In this chapter, we have presented a solution procedure for the capacitated lot-sizing and scheduling problem with setup times, setup carry-over, back-orders and parallel machines. To our knowledge, it is the first procedure able to solve such problems. The problem can be modeled as a Capacitated Lot-Sizing Problem with Linked lot-sizes, Back-Orders and Parallel Machines (CLSPL-BOPM). An in-depth literature review on big bucket lot-sizing models with respect to setup carry-over, back-ordering and parallel machines has been given.

The core of the solution procedures is a new 'heuristic model'. The basic idea is to replace the binary variables of standard formulations with integer variables, making the model solvable using standard procedures. There are two premises inherent to the model: The production volume is produced with as few setups as possible and a setup carry-over is only possible when a product's production volume (including 'dummy production volume') allows the complete utilization of a machine. Embedded in a period-by-period heuristic, the model is solved to optimality or near-optimality in every iteration using standard procedures (CPLEX) and a subsequent scheduling routine loads and sequences the products on the available machines.

Six variants of the procedure have been presented. They have been compared with a direct implementation of the CLSPL-BOPM, which is solved using a time-limited standard procedure (CPLEX). Extensive computational tests indicate that the computation time of the heuristics depends on the number of products and periods considered. The number of machines does not substantially affect the run-time. The run-time is also influenced by various problem characteristics, mainly by the relative level of inventory, back-order and setup costs as well as by capacity utilization and setup times. The effect of the problem characteristics on solution quality—i.e. the costs of a generated solution—depends on the problem size and the respective procedure. The heuristics outperform the direct CLSPL-BOPM implementation. They find better solutions in a shorter computation time. It is possible to improve the solution quality of the heuristics by allowing more runs with adjusted capacity. However, this comes at the cost of longer computation times.

In general, the heuristics solve substantially more and larger problems than the direct CLSPL-BOPM implementation in a reasonable amount of time. They are able to solve practically sized problems in a moderate amount of time and can thus be used in real business applications. As the procedures are quite general, they may be extended to incorporate various special characteristics of real world problems.

Embedded in the solution approach for flexible flow lines, the procedure solves the lot-sizing and scheduling problem for the bottleneck stage. In the subsequent phase, the schedule will be rolled out to the non-bottleneck production stages.

5

Phase II: Schedule Roll-Out

Too hot to handle?

Phase II is an intermediate phase to roll out the product family schedule from the bottleneck stage to the other stages in order to prepare the scheduling of individual products for all stages in Phase III.

Phase I of the solution procedure has covered the bottleneck stage. It has determined production quantities for all product families in each period. Phase II considers all other production stages.

Section 5.1 defines the premises and scope of Phase II. Section 5.2 presents a procedure to roll out the schedule from the bottleneck stages to the other stages. The result of the procedure are so-called 'machine/time slots' that are used as an input for Phase III. Section 5.3 points out two remarks on the resulting schedule and Sect. 5.4 gives a summary of the chapter.

5.1 Premises and Scope

For each period, the non-bottleneck stages will produce the same product family volume as the bottleneck stage in that period. Thus, the product family volumes are predetermined by Phase I, and the schedule roll-out can be performed for each period independently. In order to do so, the discrete period segmentation of the original problem is dropped and the problem is treated continuously. This is done because the original period boundaries may not allow to produce the given production volumes on all subsequent stages within a single period. However, as long as each stage does not need more than a period's time, this does not tamper with a feasible schedule. Figure 5.1 shows this schematically. The original period boundaries are plotted on the horizontal axis. The dark gray bars depict the production volume of a single period as given by Phase I. Job C is started in period 1 and finished in period 3. However, each stage produces the complete volume within an original period's length. Production volumes of the next original period from Phase I (depicted by the bright gray bars) can be produced after the volume of the actual period is completed. The whole plan is stretched, comparable to an accordion.

Phase II has a time-frame of a single period of the original lot-sizing and scheduling problem. Since the production quantities are predetermined and unchangeable, lot-sizing decisions are not made. Thus, Phase II has no impact on inventory and back-order costs of end-products. However, batching decisions must be made: We have to decide how many units to consecutively produce on one machine and how many machines to set up for each product family. The objective is to minimize the average flow time for each product family unit. Flow time is measured from the beginning of production on the first stage until the end of production on the last stage. By minimizing the average flow time, the number of intermediate product units and their share of the inventory costs is minimized.

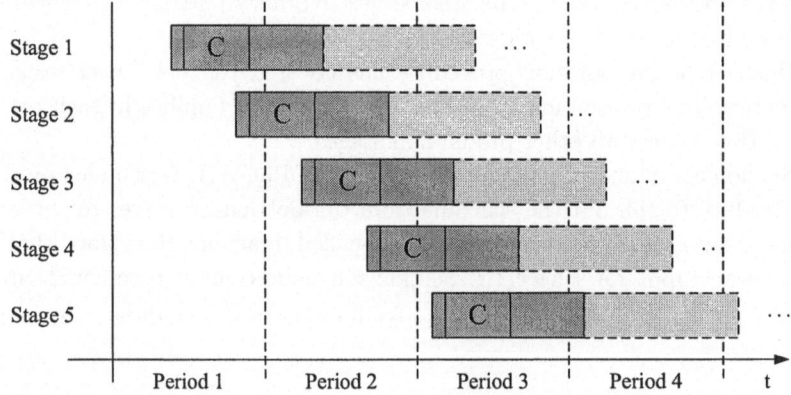

Fig. 5.1. Schedule for all production stages stretched over several original period boundaries

5.2 Scheduling Non-Bottleneck Production Stages

The basic idea of Phase II is to establish the same product family throughput on all stages (see Fig. 5.2). In this way, each preceding stage supplies the intermediate products just in time for the subsequent stage. Thus, the products do not have to wait between stages and the flow time as well as the inventory volume of intermediate products is minimized.

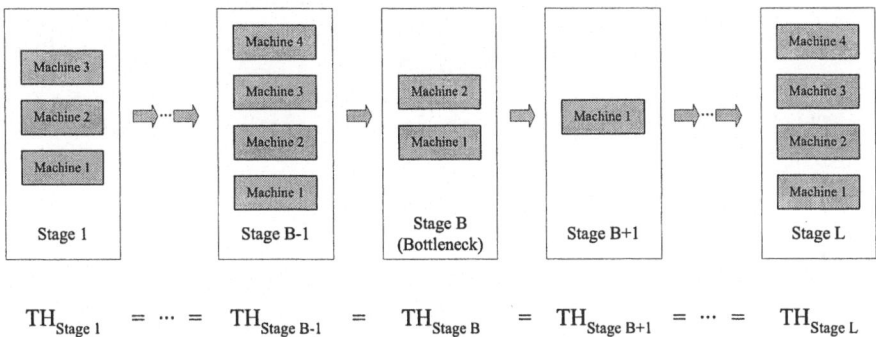

$$TH_{\text{Stage 1}} = \cdots = TH_{\text{Stage B-1}} = TH_{\text{Stage B}} = TH_{\text{Stage B+1}} = \cdots = TH_{\text{Stage L}}$$

Fig. 5.2. Schedule roll-out: Equal throughputs on all stages

The schedule roll-out consists of two steps. Step 1 determines how many and which machines to load for the different product families on the non-bottleneck stages (batching and machine-assignment). This step assumes that a job can simultaneously be produced on all production stages. However, in reality, a job has to be finished on a preceding stage before its production can start on a subsequent stage. This is incorporated in the second step, where feasible machine/time slots over all production stages are calculated. The machine/time slots indicate on which machine and at what time to produce a product family unit.

5.2.1 Step 1: Batching and Machine-Assignment

This step determines how many machines to set up on the non-bottleneck stages in order to obtain the same throughput as the bottleneck stage. The throughput on the bottleneck stage is given by the result of Phase I.

Producing a unit of product family p on the bottleneck stage takes t_p^u time units. In the remainder, the process and setup time parameters from Phase I will be augmented by a production stage index. Thus, with B as the bottleneck stage, we have $t_{Bp}^u := t_p^u$ and $t_{Bp}^s := t_p^s$ for all product families p. Suppose that, at a certain time, product family p is produced by n_{Bp} machines on the bottleneck stage. Then, the production rate for product family p can be calculated by

$$r_p := \frac{n_{Bp}}{t_{Bp}^u} .$$ (5.1)

On a non-bottleneck stage l, a single machine can produce with a production rate of up to $1/t_{lp}^u$. Let n_{lp} denote the number of parallel machines producing product family p on stage l. The production rate on stage l and B are equal if the following equation holds true:

$$\frac{n_{lp}}{t_{lp}^u} = \frac{n_{Bp}}{t_{Bp}^u}$$ (5.2)

This leads to the conclusion that the number of parallel machines on stage l must be set to

$$n_{lp} := n_{Bp} \cdot \frac{t_{lp}^u}{t_{Bp}^u} .$$ (5.3)

Calculating the number of parallel machines in this way may lead to a fractional number. Since the production rate on the bottleneck stage must be reached, $\lceil n_{lp} \rceil$ (the closest integer number larger than or equal to n_{lp}) machines must be set up for p. $\lfloor n_{lp} \rfloor$ (the closest integer number smaller than or equal to n_{lp}) of these machines produce with full utilization, i.e. a subsequent product family unit is loaded immediately after the previous one. The last machine m produces with utilization $u_{pm} := n_{lp} - \lfloor n_{lp} \rfloor$. In order to do so, the machine remains idle between each two consecutive product family units.

Let t^i denote this idle time. It can be calculated as follows: The utilization u_{pm} is the fraction of the processing time divided by the complete time for processing and idling:

$$u_{pm} = \frac{t_{lp}^u}{t_{Sp}^u + t^i}$$ (5.4)

From (5.4), it follows that the idle time t^i between each two consecutive product family units must be set to

$$t^i := \frac{t_{lp}^u}{n_{lp} - \lfloor n_{lp} \rfloor} - t_{lp}^u .$$ (5.5)

An alternative way would be to produce with the same utilization of $n_{lp}/\lceil n_{lp} \rceil$ on all machines. The idle time between two units (on all such machines) could be calculated analogously.

Now we know how many machines to set up for each product family on all stages in order to obtain the same throughput as on the bottleneck stage. However, the throughput on the bottleneck stage may change over time. Thus, the calculation has to be repeated whenever the production rate on the bottleneck stage changes, i.e. whenever a machine on the bottleneck stage starts or finishes a production batch. In this way, the number of parallel machines for each product family on the non-bottleneck stages is calculated in a chronological manner from the beginning to the end of a period, always following the production rate of the bottleneck stage.

It is possible that a product family p shall be loaded on n_{lp} parallel machines, but only $n < n_{lp}$ are available because the other machines are used by other product families (or because the process time of product family p on stage l is large). In this case, only n machines are loaded. The remaining production volume is produced as soon as machines become available again.

The procedure is shown in Fig. 5.3. It makes use of an additional parameter M_l denoting the total number of parallel machines on stage l (i.e. $M_B := M$ from Phase I). The procedure is invoked for each period of Phase I consecutively.

The presented procedure dedicates a machine to a single product family for a certain amount of time. If the machine does not produce with full utilization, idle time is included. This allows product family units to be produced at exactly the right time when they are needed on subsequent stages. However, the machines have left-over capacity. This may imply that no feasible schedule can be found—i.e. there is remaining unscheduled volume at the end of the period. The effect is stronger the fewer the relative number of machines on the current production stage. With a high number of machines, loading a machine with low utilization does not have a strong impact. With few machines, each individual machine that is not fully loaded reduces the available capacity at a higher proportion.

A solution to this problem is to dedicate a machine for more than one product family at a time. The machine will produce in a cycle that includes a number of units for each of the assigned product families. Figure 5.4 shows an example. Machine 1 produces product family a. Machine 3 produces product family b. On the bottleneck stage, both products have a production rate that cannot be reached with a single machine on the current stage. Thus, they are additionally loaded on machine 2, together with product family c, which utilizes less than a full machine on this production stage.

Cyclic production increases the utilization of the partially utilized machines. The number of machines that are needed to produce the product families decreases and a feasible schedule can be computed. However, this comes at the cost of additional setups. A setup has to be performed for each product family and cycle. Moreover, a job's flow time is prolonged as it may not be produced at exactly the time it is available from a preceding stage.

- Build chronological sequence Seq of production rate changes for all product families on bottleneck stage
- Initialize the number of available parallel machines on each stage l: Set $a_l := M_l$
- For each production rate change in Seq (let p be the affected product family, r_p its new production rate and n_{lp} the number of machines currently producing product family p on stage l):

  For each stage l other than the bottleneck stage
  - (Temporarily) release machines currently producing product family p: Set $a_l := a_l + n_{lp}$
  - Calculate number of parallel machines needed to produce production rate r_p of product family p on stage l: Set $n := r_p \cdot t_{lp}^u$
  - If $\lceil n \rceil$ machines are currently available (i.e. $\lceil n \rceil \leq a_l$)
    a) Load $\lfloor n \rfloor$ machines with product family p, produce with utilization of 100% (i.e. produce next unit as soon as previous one is finished)
    b) If $\lfloor n \rfloor \neq \lceil n \rceil$: Load an additional machine with product p, produce with utilization $n - \lfloor n \rfloor$ (i.e. insert an idle time of $\frac{t_{lp}^u}{n - \lfloor n \rfloor} - t_{lp}^u$ between each two units)
    c) Update remaining number of available machines: Set $a_l := a_l - \lceil n \rceil$

    else
    a) Load all remaining available machines a_l with product p, produce with utilization of 100% (i.e. produce next unit as soon as previous one is finished)
    b) Load leftover $\lceil n \rceil - a_l$ machines as soon as they become available again
    c) Update remaining number of available machines: Set $a_l := 0$

Fig. 5.3. Scheduling stages other than the bottleneck stage

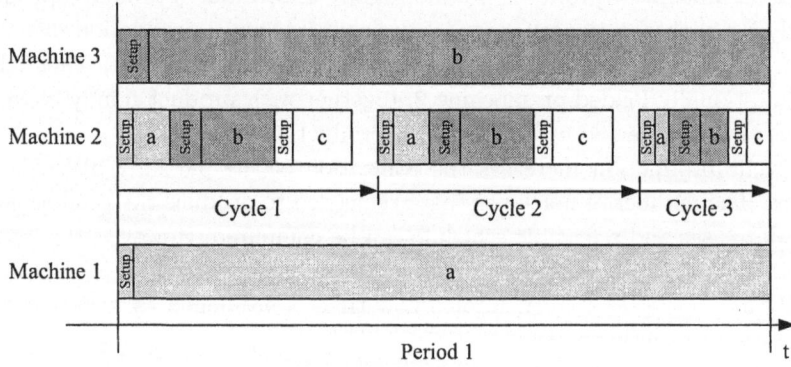

Fig. 5.4. Cyclic production

An important parameter is the cycle time. The cycle time is measured by the time between two setups (or batches) for the same product family. It is indicated by the arrows in Fig. 5.4. The cycle time determines the number of units produced in each cycle. With a cycle time of CT_m and a designated production rate of r_{pm} for a product family p on the current machine m, the number of units per batch is $b_{pm} := r_{pm} \cdot CT_m$. As this may be a fractional number, it is rounded up or down in each cycle, depending on whether it is under- or overreached at the moment. Every product family is produced once per cycle.

With a relatively short cycle time, the machine will perform setups for relatively small product family batches. Thus, the shorter the cycle time, the higher the number of setups. On the other hand, the longer the cycle time, the worse the flow time because a product family is produced only once per cycle. For that reason, a minimum and a maximum cycle time have to be provided externally.

The maximum cycle time has an impact on the number of product families that can be incorporated in a cycle: Each product family p uses a fraction of size u_{pm} of the machine's capacity for processing. Let P_m^c denote the product families that are supposed to share a single machine m. The sum of their utilizations cannot be greater than one:

$$\sum_{p \in P_m^c} u_{pm} \leq 1 \tag{5.6}$$

The remaining cycle time can be used for setups. It must be long enough to allow a setup for each product family of P_m^c:

$$\sum_{p \in P_m^c} t_{lp}^s \leq CT_m \left(1 - \sum_{p \in P_m^c} u_{pm} \right) \tag{5.7}$$

The left hand side computes the sum of all setup times which must be less than or equal to the remaining cycle time as calculated on the right hand side. With a given maximum cycle time of $CT^{\max}$, it follows that condition (5.8) must hold true:

$$\frac{\sum_{p \in P_m^c} t_{lp}^s}{1 - \sum_{p \in P_m^c} u_{pm}} \leq CT^{\max} \tag{5.8}$$

Thus, a new product family can only be added to an existing cycle if the total utilization will not exceed 100% and the given maximum cycle time will not be exceeded. The longer the maximum cycle time, the more product families may be loaded on a single machine.

With a given set of product families to share a machine, the shortest possible cycle time CT^s can be calculated using formula (5.9):

$$CT^s := \frac{\sum_{p \in P_m^c} t_{lp}^s}{1 - \sum_{p \in P_m^c} u_{pm}} \tag{5.9}$$

However, CT^s may be relatively small and result in a high number of setups. For that reason, a minimum cycle time $CT^{\min}$ must be given externally and will be used if it is smaller than CT^s:

$$CT_m := \begin{cases} CT^s & \text{if } CT^s \geq CT^{\min} \\ CT^{\min} & \text{else} \end{cases} \tag{5.10}$$

If CT_m is greater than CT^s, some idle time will be added at the end of each cycle as in the case of Fig. 5.4. The last cycle may be shorter than CT_m if the period boundary is reached or if the machine assignment changes. In this case, the batch-sizes are reduced proportionally (see cycle 3 in Fig. 5.4).

The schedule roll-out procedure from Fig. 5.3 can be augmented by cyclic production by replacing step b of the 'if-case' with the sub-procedure shown in Fig. 5.5. At any given time, there will be, at most, one machine for each product family and production stage that produces in a cyclic way. This machine has to be retrieved whenever a product family's production rate changes. If there is a machine that has been producing the product family in a cyclic way until the production rate change, it will be used with priority. Thus, we check if the product family can be reloaded on this machine. The production rate change implies a changed utilization on the cycling machine. If the utilization caused by the current product family is higher than before, the machine might not be able to produce the product family any longer. In this case, we check all other machines producing in a cycle whether they can additionally load the current product family. The sequence in which those machines are considered may be chosen arbitrarily. If none of these machines is able to produce the product family, it is assigned to a new machine that is unused at the moment. This machine may later be loaded by other product families if the utilization and cycle time restrictions allow.

5.2.2 Step 2: Creating a Feasible Product Family Schedule – Generation of Machine/Time Slots

Result of Phase II (Schedule roll-out) is a feasible product family schedule that consists of production slots for all product family units on all stages. These slots will be called 'machine/time slots' in the remainder and indicate the machine and production time of a product family unit.

1. If a machine m has been producing product family p in a cyclic way until the current production rate change (note that $p \notin P_m^c$ because p has been temporarily unloaded)

   If $u_{pm}+\sum_{q \in P_m^c} u_{qm} \leq 1$ and $CT^s := \dfrac{t_{lp}^s+\sum_{q \in P_m^c} t_{lq}^s}{1-\left(u_{pm}+\sum_{q \in P_m^c} u_{qm}\right)} \leq CT^{\max}$

   - Load product family p on machine m:
     Set $P_m^c := P_m^c \cup \{p\}$, cycle time $CT_m := \min \{CT^s, CT^{\min}\}$ and a production quantity of $b_{qm} := r_{qm} \cdot CT_m$ units per cycle for each product family $q \in P_m^c$. In each cycle, round up b_{qm} to the next greater integer if the average batch-size of earlier cycles is smaller than b_{qm}. Round down otherwise.
   - Exit this sub-procedure
2. For all machines m producing product families in a cyclic way

   If $u_{pm}+\sum_{q \in P_m^c} u_{qm} \leq 1$ and $CT^s := \dfrac{t_{lp}^s+\sum_{q \in P_m^c} t_{lq}^s}{1-\left(u_{pm}+\sum_{q \in P_m^c} u_{qm}\right)} \leq CT^{\max}$

   - Load product family p on machine m:
     Set $P_m^c := P_m^c \cup \{p\}$, cycle time $CT_m := \min \{CT^s, CT^{\min}\}$ and a production quantity of $b_{qm} := r_{qm} \cdot CT_m$ units per cycle for each product family $q \in P_m^c$. In each cycle, round up b_{qm} to the next greater integer if the average batch-size of earlier cycles is smaller than b_{qm}. Round down otherwise.
   - Exit this sub-procedure
3. Select a currently empty machine m and load product family p on machine m: Set $P_m^c := \{p\}$. Produce with utilization u_{pm} (i.e. insert an idle time of $\frac{t_{lp}^u}{u_{pm}} - t_{lp}^u$ between each two units, analogously to step b of the 'if-case' of Fig. 5.3) until other product families are added to the cycle.

Fig. 5.5. Sub-procedure to load machines with cyclic production (replaces step b of the 'if-case' in Fig. 5.3)

In step 1 of Phase II, production of a product family unit starts simultaneously on all stages. However, in reality, a job has to be finished on a preceding stage before it can be processed on a subsequent stage. The job start-times have to take these precedence constraints into account.

The plan for the first production stage can be used as given by the procedure from step 1. There is no earlier stage that may postpone a job's start-time. To create a coordinated schedule for stage 2, we build a chronological sequence of all jobs and their associated machines on stage 1 and 2 as scheduled by step 1. The sequences include the jobs of all product families. For each job j on stage 2 in this order, we locate the first yet unused corresponding job—i.e. a job representing the same product family as job j—in the sequence

---

For all stages l from 1 to $L - 1$

/* Schedule stage $l + 1$ */

- Build sequence Seq^l of chronologically sorted jobs and their associated machines for all product families on stage l
- Build sequence Seq^{l+1} of chronologically sorted jobs and their associated machines for all product families on stage $l + 1$
- For all jobs j of Seq^{l+1} (let p be the product family of job j and m the associated machine. Let t_m be the time machine m becomes idle):
  1. If machine m is not set up for product family p
     - Set up machine m for product family p, starting at time t_m
     - Update the time machine m becomes idle:
       Set $t_m := t_m + t^s_{l+1,p}$
  2. Locate the first yet unused corresponding job (same product family p) in Seq^l. Let t be the time it is finished on stage l
  3. Fix the start-time of job j on machine m to the minimum of t and t_m.
  4. Update the time machine m becomes idle: Set $t_m := t_m + t^u_{l+1,p}$

Fig. 5.6. Generation of product family machine/time slots for all production stages

of stage 1. Production start- and end-times on stage 1 are fixed. Production on stage 2 will now be fixed to the earliest possible time: If the machine has to perform a setup, this can be done as soon as the machine becomes idle after processing a possible previous job. Processing of job j will start after the machine on stage 2 is ready and the corresponding job on stage 1 is finished, whichever occurs later. After producing the job, the machine becomes idle again and can load the next job. Having done this for all jobs on stage 2, the final schedule for stage 3 can be created using the same procedure: At this point, the schedule for stage 2 is fixed and can be used as input for stage 3. In this way, all subsequent stages are scheduled iteratively until a schedule for the last stage L has been generated. Figure 5.6 explains the procedure in detail. The procedure preserves the sequence of jobs on a machine as determined by step 1.

5.3 Two Remarks on the Resulting Schedule

The first remark concerns the setup costs used in Phase I and the number of setups on the non-bottleneck stages: The number of parallel machines producing a product family on the bottleneck stage determines the throughput

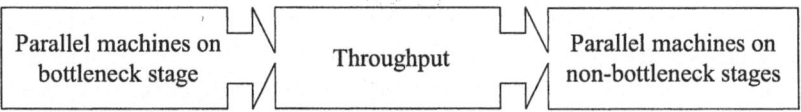

Fig. 5.7. Connection between the number of parallel machines and the throughput

of the product family. In turn, the throughput specifies the number of parallel machines on the non-bottleneck stages (see Fig. 5.7). Thus—with the exception of cyclic production—each setup on the bottleneck stage incurs a fixed number of setups on the other stages. Hence, it is possible to calculate system-wide setup costs incurred by a setup on the bottleneck stage. These setup costs can be used in Phase I. Moreover, the per period production volume of a product family on the bottleneck stage is produced with the minimal number of machines possible. Hence, the behavior of Phase I (to use as few machines as possible for a product family) implies that the number of parallel machines (and setups) on the non-bottleneck stages is limited as well.

The second remark is on job waiting and machine idle times: Having established the same throughput on all stages does not prevent some units from having to wait in the intermediate buffers. This is due to the discrete nature of the problem. When a product unit is finished on a preceding stage, it might have to wait for a machine on the subsequent stage to finish its actual unit. Figure 5.8 shows a small example with two machines on stage 1 delivering

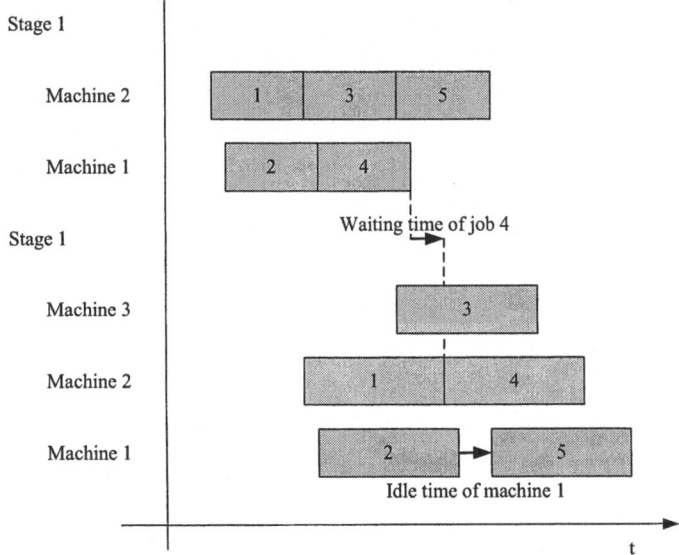

Fig. 5.8. Waiting times of jobs and idle times of machines

input for three machines on stage 2. Job 4 has to wait until machine 2 on stage 2 has finished job 1. The figure also shows that a machine may be idle between jobs, even if the assigned product family was supposed to completely fill the machine: Machine 3 on stage 2 waits for job 5. However, these waiting and idle times are relatively short.

5.4 Summary

Phase II rolls out the product family schedule from the bottleneck stage to the other production stages. The objective is to minimize mean flow time. This is accomplished by establishing the same product family throughput on all stages. In this way, a job's waiting time between stages and hence its flow time through the system is minimized.

The result of Phase II are so-called 'machine/time slots' that indicate on which machine and at what time to produce a product family unit. The subsequent planning phase will decide which product of the respective product family to assign to each of these slots. Thus, the flow time on product level will be determined in Phase III. Therefore, computational tests of Phase II will be performed together with the ones of Phase III and presented in Sect. 6.4.

6

Phase III: Product-to-Slot Assignment

Deal with the devil at your own risk!

The previous phase has created time and machine windows for each product family on all production stages. In this phase, we have to disaggregate the product families to individual products and solve the associated scheduling problem: For each product family and production stage, we have a number of product family unit production start- and end-times and associated machines. The problem is to assign a product unit to each of those product family slots. In other words, we have to label all the existing product family machine/time slots with a product name.

Section 6.1 describes the objectives of this phase and points out that there is a trade-off between them. Section 6.2 introduces Genetic Algorithms and their components. Section 6.3 presents a solution procedure based on two nested Genetic Algorithms. Computational results are provided in Sect. 6.4. We summarize the chapter in Sect. 6.5.

6.1 Goal Trade-Off

The objective in this phase is to minimize intra-family setup costs while trying to keep the low flow time from Phase II. We assume that on any production stage, all products of a family share the same setup costs. Thus, minimizing a family's intra-family setups costs is equivalent to minimizing the number of intra-family setups, weighted by the setup costs on the respective production stage.

Figure 6.1 shows a typical situation for two consecutive stages. In this example, we have an excerpt of the product family schedule for a product family C, consisting of products $C1$ and $C2$ with a demand of 10 respectively 8 units. The initial situation after the schedule roll-out is depicted in panel a of Fig. 6.1. On stage l, there is one machine and on stage $l+1$, there are four machines scheduled for product family C. Looking at stage $l+1$ first and trying to minimize the number of setups, we would like to load the first two machines with product $C1$ and the last two machines with product $C2$ (panel b of Fig. 6.1), leading to only four setups on stage $l+1$, which is also the minimum number of setups we have to perform. In order to achieve a low flow time, stage l has to produce the units in the same order as they are needed on stage $l+1$. This leads to a very high number of setups on stage l. On the other hand, considering stage l first—and scheduling all units of $C1$ before performing a setup to $C2$—leads to eight setups on stage $l+1$, as all four machines have to produce $C1$ and then change to $C2$ (panel c). Trying to perform as few setups as possible, we get a plan as shown in panel d of Fig. 6.1: Since two machines on stage $l+1$ cannot reach the production rate of stage l, we build up a work in process inventory for product $C1$ and $C2$

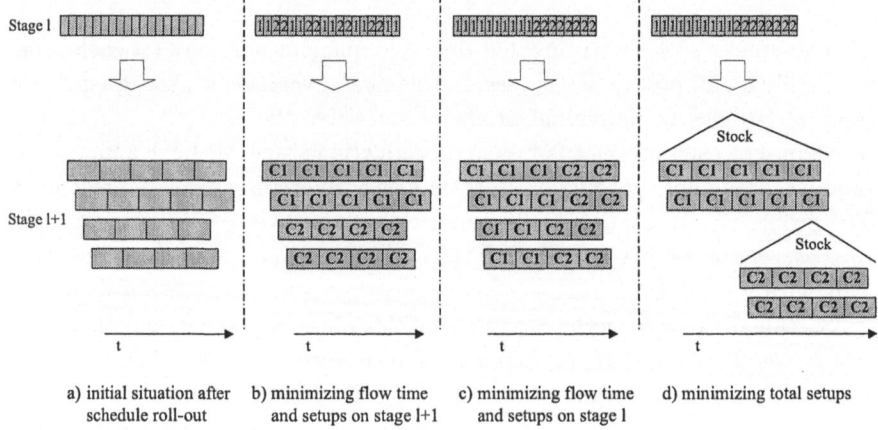

a) initial situation after   b) minimizing flow time   c) minimizing flow time   d) minimizing total setups
schedule roll-out   and setups on stage l+1   and setups on stage l

Fig. 6.1. Trade-off between number of setups and flow time

between stages l and $l + 1$. Moreover, the last two machines of stage $l + 1$ cannot start at the pre-determined time because at that time no unit of $C2$ is ready for production yet.

To sum up, there is a trade-off between minimizing the number of setups (i.e. setup costs) and the flow time and we have to find solutions for the following four sub-problems: For all stages, we have to determine (1) how many machines to set up for each specific product, (2) in which batch-size, (3) when in the given time window and (4) on which of the pre-determined machines to produce the product units. We develop a solution procedure that consists of two nested Genetic Algorithms to solve these problems.

6.2 Genetic Algorithms

Genetic Algorithms were developed in the 1960s (see e.g. Fogel et al. 1966 or Holland 1975). They represent a class of algorithms that imitate natural behavior—the evolutionary process—in order to solve difficult optimization problems.

A Genetic Algorithm maintains a set of candidate solutions. Each solution is called an 'individual'. The set of solutions is called 'population'. With each iteration, a new population (or 'generation') is formed by selecting two individuals and mating them. This leads to a new, 'offspring' solution. Analogously to the natural notation, this process is called 'crossover'. The individuals to enter the crossover are selected by their 'fitness' value. The fitness value is derived from the objectives of the original optimization problem. The better a solution, the higher the individual's fitness value. Thus, with the selection-operator, a Genetic Algorithm incorporates the concept of 'survival of the fittest' into optimization problems. The idea is that, over the generations, better solutions will evolve by repeatedly combining and restructuring parts of good solutions. Ultimately, an optimal or at least good solution will be found.

In order to overcome local optima, another operator, called 'mutation', is included. It modifies a single individual. It has a diversification effect as it is able to create completely new and different solutions. The crossover, on the other hand, intensifies the search in promising regions of the search space.

Genetic Algorithms start with an initial population of solutions. These solutions may be created randomly or by other (simple) procedures. To be able to employ the crossover- and mutation-operators, each individual must be encoded in a special structure, the 'representation format'. Early Genetic Algorithms used simple binary coding mechanisms that represent solutions by 0-1 bit strings. Later, more sophisticated representation formats were developed. Consider a single machine scheduling problem as an example: A

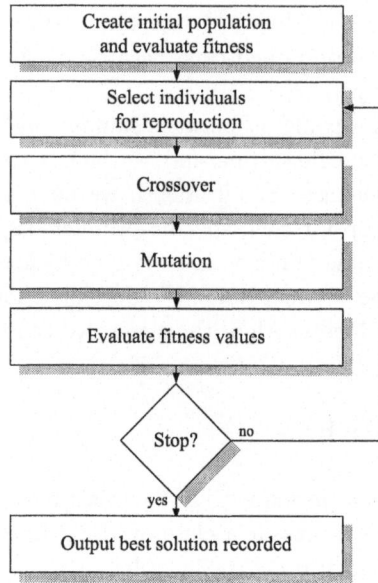

Fig. 6.2. General structure of a Genetic Algorithm

possible representation format for a solution is a string of integers, indicating the job numbers in the order in which they are processed on the machine. Multi-machine environments require more elaborate representation formats.

There must be an unambiguous function that transforms the encoded individual to a solution of the original problem. This function is used for the output of the best solution at the end of the Genetic Algorithm. Moreover, in most cases, an individual's fitness cannot be assessed in its encoded format. Therefore, it must be possible to compute this function efficiently as it has to be applied to every new individual.

Since a Genetic Algorithm is not able to verify if it has found an optimal solution, we have to use a special stopping criterion. A simple criterion is to stop after a certain number of generations. Another criterion would be to stop after a certain number of consecutive generations without an improvement.

Figure 6.2 summarizes the general structure of a Genetic Algorithm.

A detailed introduction to Genetic Algorithms is given by Michalewicz and Fogel (2000) or Reeves and Rowe (2003). Genetic Algorithms have been used for very diversified fields of research, such as biology, engineering or social sciences. Goldberg (2003) gives an overview. Genetic Algorithms have also proven successful for many production and operations management problems. Recent overviews are given by Dimopoulos and Zalzala (2000) and Aytug et al. (2003). We present two nested Genetic Algorithms for Phase III of the solution approach.

6.3 Solution Procedure: Two Nested Genetic Algorithms

Phase II has generated a plan for every product family. Thus, we can solve the scheduling problem for each of the product families independently. The stage with the largest number of parallel machines (stage $l+1$ in the example of Fig. 6.1) appears to be the most difficult stage. Therefore, it is considered first and will be called stage S hereafter.

The scheduling problem is solved by a hybrid procedure consisting of two nested Genetic Algorithms, which is invoked for each product family separately. We begin solving the above mentioned sub-problems (1) and (2) by setting target values for the number of machines and the batch-size for each product of the actual family on stage S. The corresponding procedure is explained in Sect. 6.3.1. The target values are handed over to the next step, where a schedule is created by solving sub-problems (3) and (4), i.e. by fixing the exact production time and machine for each product unit. This is done by means of an inner Genetic Algorithm. Its basis is an ordered list (sequence) that includes one entry for each product of the actual family. The inner Genetic Algorithm transforms such a sequence into a schedule for stage S. This schedule delivers a chronological sequence of all product units, which is used to assign the machine/time slots on the other stages. However, some local re-sequencing is allowed as long as a job's flow time through the stages is not increased above a certain limit. The inner Genetic Algorithm is described in Sect. 6.3.2. In its last step, the inner Genetic Algorithm calculates the intra-family setup costs to evaluate the current solution. Based on this evaluation, new target values are generated by an outer Genetic Algorithm presented in Sect. 6.3.3. Thus, an iterative process is invoked that stops when no improvement can be made within a certain number of iterations or when a maximum number of iterations is reached. Figure 6.3 gives a summary of the solution procedure.

According to a classification scheme introduced by Talbi (2002), the two nested Genetic Algorithms form a hybrid metaheuristic. Using the suggested notation, its scheme can be written as *LTH(GA(GA))(homogeneous, global, specialist)*. The expression LTH(GA(GA)) implies a 'Low-level Teamwork Hybrid' consisting of two Genetic Algorithms. 'Low-level' means that a function of the outer procedure—the evaluation of the target values of the outer Genetic Algorithm—is performed by another procedure. 'Teamwork' means that there are parallel cooperating agents in the outer procedure, which is always true for a population based approach. The fact that both parts employ the same metaheuristic—a Genetic Algorithm—makes the procedure 'homogeneous'. The inner and the outer Genetic Algorithm perform different tasks—one sets target values, the other one creates a schedule. Hence, they are 'specialists'. 'Global' means that each part searches through its complete solution space.

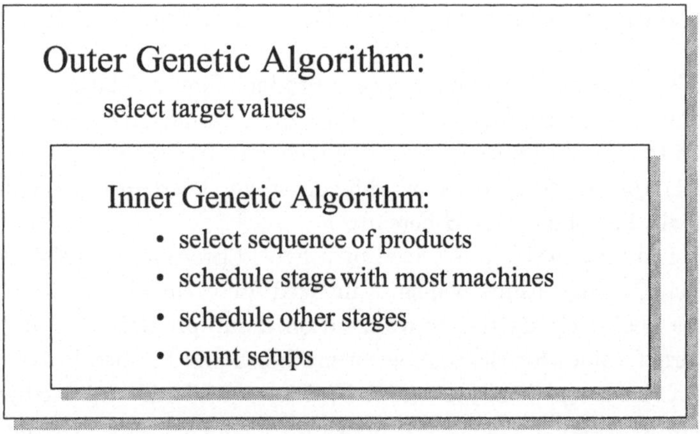

Outer Genetic Algorithm:
select target values

Inner Genetic Algorithm:
- select sequence of products
- schedule stage with most machines
- schedule other stages
- count setups

Fig. 6.3. Overview of the product scheduling procedure

6.3.1 Setting Target Values

This step considers the stage with the largest number of parallel machines (stage S). The objective is to calculate a target number of setups (i.e. a target number of parallel machines) for each product of the current product family. A product's target batch-size follows by dividing its demand by the target number of setups.

The calculation is based on a target number of setups for the complete product family. This figure is lower- and upper-bounded: For every product of the family, at least one setup has to be made. Thus, the total number of setups per product family cannot be less than the number of products contained in the family. Therefore, the number of setups for product family p on stage S is limited by the lower bound

$$S_p^{LB} := K_p \, , \tag{6.1}$$

where K_p is the number of products in product family p. A tighter lower bound would include the number of machines scheduled for the product family. For each of those machines, at least one setup has to be performed. This would lead to an alternative lower bound of $S_p^{LB2} := \max \left\{ M^{Sp}, K_p \right\}$, where M^{Sp} is the total number of machines for product family p on stage S in the schedule retrieved from Phase II. However, the objective here is not to derive tight bounds, but to restrict the possible solutions for the outer Genetic Algorithm. The final number of setups may be different from the target value. Target values below S_p^{LB2} (but greater than S_p^{LB}) may lead to different and possibly better solutions than target values above S_p^{LB2}. Hence, they should be considered as well.

For an upper bound, we assume that a machine produces all jobs of a product consecutively. The smallest batch-size is a single job. If all jobs of a product are placed on different machines, the maximum number of setups per product cannot be larger than the number of machines for the product family or the demand in jobs. This means that the number of setups for product family p on stage S is limited by the upper bound

$$S_p^{UB} := \sum_{k=1}^{K_p} \min \left\{ M^{Sp}, D_{pk} \right\} , \tag{6.2}$$

where D_{pk} is the demand for product k of product family p and the other symbols as above. The demand volumes are derived from the scheduled production volumes from Phase I: The demand for the complete product family is set to $D_p := \sum_{m \in \bar{M}} x_{ptm}$. It has to be distributed to the individual products so that $D_p = \sum_{k=1}^{K_p} D_{pk}$ holds true. As the inventory costs within a product family are identical, this can be done by a simple chronological matching of the externally given, original product demands and the scheduled production volumes over the periods.

By means of an outer Genetic Algorithm described in Sect. 6.3.3, a target number of setups (T_p^s) for each product family p is chosen from the interval $\left[S_p^{LB}; S_p^{UB} \right]$ and an associated target batch-size (T_p^b) (which might be a fractional number) is calculated by dividing the product family's demand by that figure:

$$T_p^b := \frac{D_p}{T_p^s} \tag{6.3}$$

The setups have to be distributed to the different products of the product family, i.e. we have to determine a target number of setups—or, equivalently, a target number of parallel machines—for each individual product. The goal is to generate similar batch-sizes for all products within a product family. This is achieved in the following way: First, the remaining demand $D_p^r := D_p$ for product family p is initialized with the total product family demand and the remaining number of setups for the product family $S_p^r := T_p^s$ is initialized with the target number of setups. Then, the products of family p are ordered by non-decreasing demand volume. If the demand of a product k is lower than or equal to the family's actual target batch-size, the target number of setups for the product (T_{pk}^s) is set to one machine. If the demand is higher than the family's target batch-size, the number of setups for the product is calculated by dividing its demand by the family's target batch-size and setting it to the nearest integer number to the result. After each product, the remaining demand, the remaining setups, and the target batch-size for the product family are updated. Figure 6.4 summarizes the calculation of the target values: The

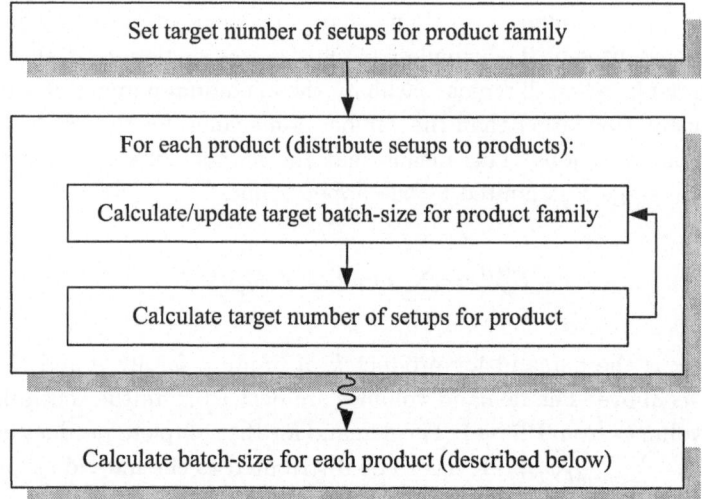

Fig. 6.4. Overview of the target value calculation

target number of setups for the product family is distributed to the individual products by means of the associated target batch-size. The target number of setups for the individual products will in turn be used to calculate the batch-sizes of the respective products, as will be described below. Figure 6.5 shows the detailed procedure.

For each product family p

1. Set target number of setups for product family p (T_p^s) by means of outer Genetic Algorithm
2. Initialize data: Set remaining product family demand $D_p^r := D_p$ and remaining product family setups $S_p^r := T_p^s$
3. For all products k of product family p in non-decreasing order of D_{pk}

   a) Calculate target batch-size for product family: Set $T_p^b := \frac{D_p^r}{S_p^r}$

   b) Set target number of setups for product k:

$$
T_{pk}^s := \begin{cases} 1 & \text{if } D_{pk} \leq T_p^b \\ \text{round}\left(\frac{D_{pk}}{T_p^b}\right) & \text{if } D_{pk} > T_p^b \end{cases}
$$

   c) Update remaining product family demand: Set $D_p^r := D_p^r - D_{pk}$
   d) Update remaining product family setups: Set $S_p^r := S_p^r - T_{pk}^s$

Fig. 6.5. Algorithm to calculate the target number of setups for each product

Table 6.1. Example for obtaining the target number of setups per product

Product (k)	C1	C2	C3	C4	Sum
Demand $(D_{C,k})$	6	6	3	5	20
Rank of demand volume	3	4	1	2	
Updated target batch-size (T_C^b)	2	2	2.2	2.1	
Target number of setups $(T_{C,k}^s)$	3	3	1	2	9

Consider the following example for a product family C: Product family C consists of product $C1$ with a demand of six units, $C2$ with a demand of six units, $C3$ with a demand of three units and $C4$ with a demand of five units, totalling 20 units. The target number of setups for family C has been set to $T_C^s = 9$. Product $C3$ will be considered first, as it has the lowest demand volume. The target batch-size is set to $T_C^b = 20/9 \approx 2.2$ and the target number of setups for $C3$ is set to $T_{C,C3}^s = \text{round}(3/2.2) = 1$. T_C^b is updated to $(20 - 3)/(9 - 1) = 17/8 \approx 2.1$ and the target number of setups for $C4$ is set to $T_{C,C4}^s = \text{round}(5/2.1) = 2$. In the next iteration, $C1$ and $C2$ have the same demand volume, so we choose $C1$ arbitrarily. We set $T_C^b = (17 - 5)/(8 - 2) = 12/6 = 2$ and $T_{C,C1}^s = \text{round}(6/2) = 3$. Finally, T_C^b is updated to $(12 - 6)/(6 - 3) = 6/3 = 2$, so the target number of setups for $C2$ is set to $T_{C,C2}^s = \text{round}(6/2) = 3$. The results are shown in Table 6.1.

The target number of setups and the target batch-size for each product of the product family are solutions to sub-problems (1) and (2) from Sect. 6.1. They are handed over to the inner Genetic Algorithm.

6.3.2 The Inner Genetic Algorithm: Creating a Product Schedule Using the Target Values

The inner Genetic Algorithm generates a schedule on product level by determining when in the given product family time window and on which of the pre-determined machines to produce the jobs. These two decisions are sub-problems (3) and (4) from Sect. 6.1.

6.3.2.1 Solution Coding

A solution (i.e. an individual) of the inner Genetic Algorithm is coded in the form of a product sequence. In the above example for product family C consisting of products $C1$, $C2$, $C3$ and $C4$, one possible solution would be $(C1, C3, C2, C4)$. In general, we denote a sequence for product family p by $(\pi_1, \pi_2, \ldots, \pi_{K_p})$, where each π_i $(i = 1, 2, \ldots, K_p)$ represents one of the K_p products of the family.

6.3.2.2 Schedule Generation: A Deterministic Loading Procedure

A deterministic loading procedure is used to obtain a schedule for all production stages. A schedule consists of precise time and machine assignments for each product unit. Thus, the loading procedure has to transform a solution that is represented by a product sequence into a real schedule. This is done using the target values from Sect. 6.3.1. The procedure is called deterministic because, to a given sequence (and target values), it always returns the same schedule.

We divide the planning period into several sub-periods. This is done using an externally given parameter indicating the number of such sub-periods. For example, with a planning period of (half) a week, the sub-periods could be shifts. As mentioned in Sect. 6.3.1, the stage with the largest number of parallel machines (stage S) seems to be the most difficult stage. For that reason, we consider it first. The product family schedule from Phase II is taken as a basis. We count the number of parallel machines in each sub-period allocated to the current product family on stage S. A production slot of the product family schedule is allotted to its start-time sub-period. The deterministic loading procedure will assign an individual product to each of the machine/time-slots.

The idea is, that when we look at the situation as a Gantt-Chart, each product—with its target number of setups (i.e. the target number of parallel machines) and the corresponding target batch-size—forms a two-dimensional object that has to be packed on a pallet formed by the given machine and time window of the product family schedule. Hence, the situation resembles a two-dimensional packing problem (see e.g. Hopper and Turton 2001 or Lodi et al. 2002). Figure 6.6 illustrates the analogy. The inner Genetic Algorithm is based on a Genetic Algorithm for a two-dimensional packing problem described by Jakobs (1996).

In general, it will not be possible to find a packing pattern that covers the machine and time window exactly. Because of this, we allow the tentative target values to be overwritten in some cases: We load the products in order of the given sequence in a forward-manner starting with the first sub-period that contains a slot for the actual product family. Whenever we load the first unit of a product, we determine the number of parallel machines for the product by considering its target number of setups and the still unassigned available machines in the current sub-period only. Except for one case described below, once the number of parallel machines is set for a product, it will not be changed in later sub-periods, even if further machines become available. When a product is first considered, there are two cases: Either the number of available machines in the current sub-period is higher than or equal to the target number of setups, or it is lower than the target number of setups. In the for-

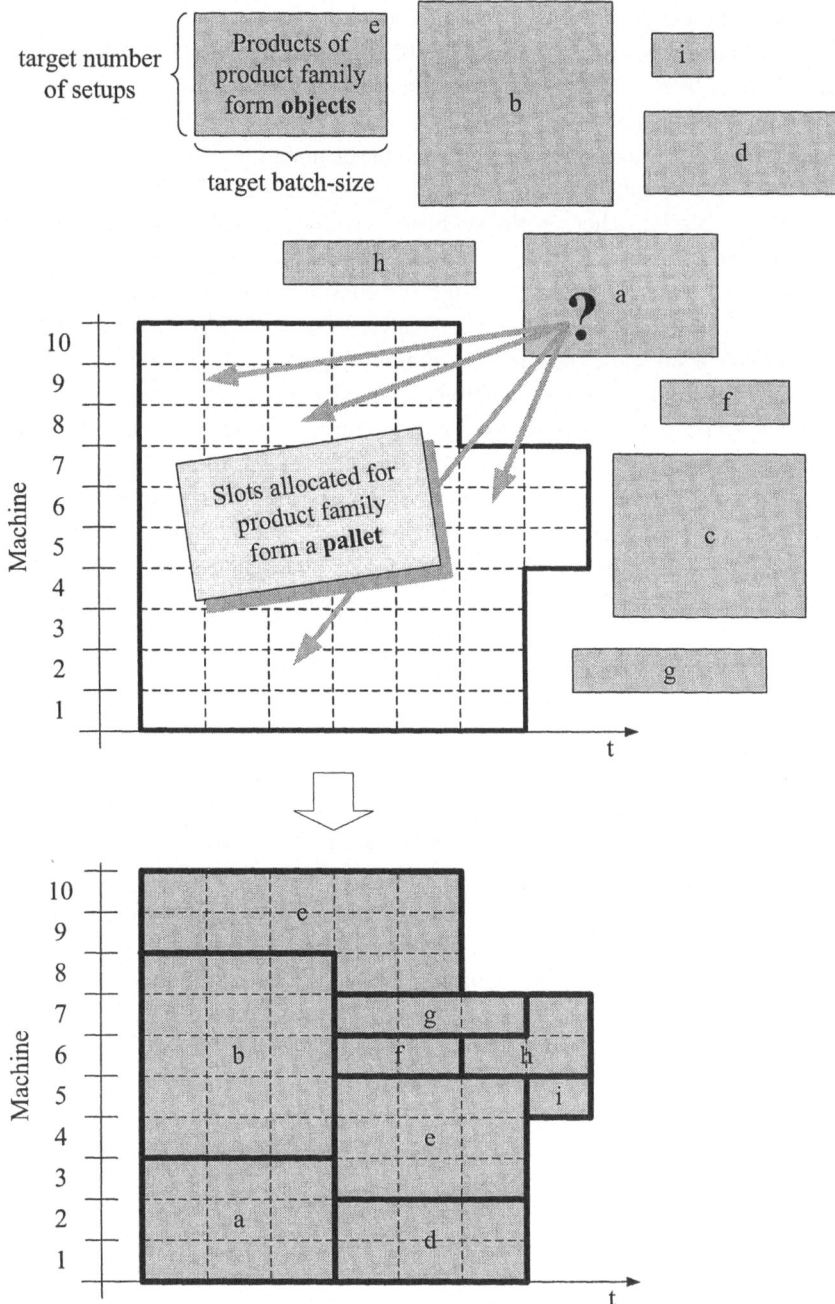

Fig. 6.6. Scheduling products on stage S resembles a two-dimensional packing problem

mer case, the target number of parallel machines is set up for the product. In the latter case, we load only the available number of machines and ignore the target value. In both cases, the objective is to divide the product units equally over the allotted machines. Obviously, in the latter case, the batch-sizes are larger than the target value as there are fewer machines to produce the units. The units are assigned to the designated production slots and the machines are unavailable for other products during production. They become available again in the sub-period of their first still unassigned production slot. If a selected machine does not have enough production capacity, i.e. the number of unassigned slots on the machine is smaller than the target batch-size, all of the machine's slots are assigned. The remaining units of a product are treated as a new product, which is scheduled immediately at the actual position in the product sequence using the same procedure. If there are no more available machines in the current sub-period, the number of parallel machines for such a product may increase in later sub-periods. As mentioned above, this is the only case where this can happen. When all parallel machines of a sub-period are scheduled, i.e. whenever there is no more unassigned production slot in a sub-period, we jump to the next sub-period with an unassigned production slot.

Figure 6.7 gives an example of the procedure. It shows the machine/time-slots allotted to product family C of the above example. The sequence to evaluate is $(C1, C3, C2, C4)$. We start with assigning slots to product $C1$. Sub-period 1 is the first sub-period with slots for product family C. There are three parallel machines in sub-period 1. Since the target number of setups for product $C1$ is three (see Table 6.1), we load all three machines with product $C1$. There are six units to schedule, so each machine produces two units. The next product in the sequence is $C3$. There are unassigned slots in sub-period 2. $C3$ has a target number of one setup, so we load it on one machine (we chose machine 3 arbitrarily, the details for the machine selection will be explained below). The units of $C3$ fill all available slots of machine 3 in sub-period 2. Product $C2$ has to be scheduled next. There are unassigned slots in sub-period 2. The target number of setups for product $C2$ is three, but the number of parallel machines with unassigned slots in sub-period 2 is only two. Thus, the target number of setups will be ignored and machines 1 and 2 will be loaded with three units each. Though it is obvious that all remaining slots must be allotted to product $C4$ because it is the last one in the sequence, we continue with the formal description of the algorithm: The next sub-period with unassigned slots is sub-period 3. Product $C4$ has to be scheduled with a target number of two setups. We load machine 1 and machine 2 with two units each, because they have the highest number of remaining unassigned slots (details below). There is a demand of one unit leftover to schedule for

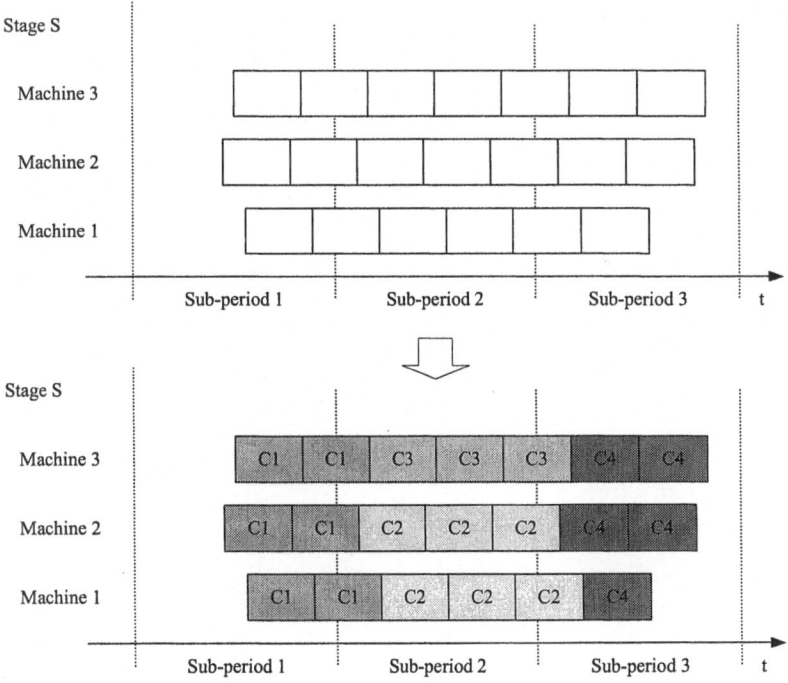

Fig. 6.7. Example of the inner Genetic Algorithm's deterministic loading procedure for stage S: Loading fewer than the target number of machines for product $C2$

product $C4$. Since there is an unassigned slot left in sub-period 3, the job is scheduled in this sub-period, filling machine 1. All in all, we have generated a schedule with nine setups, equal to the target number.

Another example is given in Fig. 6.8. A product family D consists of products $D1$, $D2$, $D3$ and $D4$. The target number of setups for the family is six. The demand volumes and resulting target number of setups for the respective products are shown in Table 6.2. Let $(D1, D3, D2, D4)$ be the sequence to evaluate. In the first sub-period, there are two parallel machines available, which coincides with the target number of setups for $D1$, the first product in the sequence. Hence, three units are loaded on machine 1 and two units on machine 2. The next product is $D3$ with a target number of one setup. The next sub-period with available slots is sub-period 2, where all three parallel machines have unassigned slots. Product $D3$ is loaded on machine 3. There are two machines left in sub-period 2 and product $D2$ is next in the sequence with a target number of two setups. Four units are loaded on machine 2. The remaining three units should be loaded on machine 1, but there are only two slots left. Thus, one unit of $D2$ remains unassigned in this iteration. The next sub-period with unassigned slots is sub-period 3, where machine 3 becomes

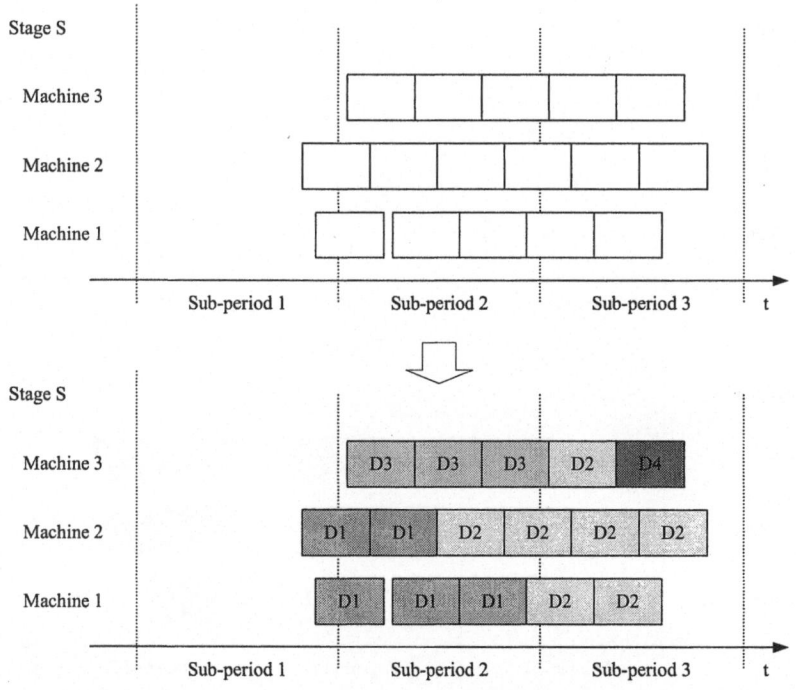

Fig. 6.8. Example of the inner Genetic Algorithm's deterministic loading procedure for stage S: Loading less than the target batch-size of product $D2$

Table 6.2. Data for a second example of the inner Genetic Algorithm's deterministic loading procedure

Product (k)	$D1$	$D2$	$D3$	$D4$	Sum
Demand $(D_{D,k})$	5	7	3	1	16
Rank of demand volume	3	4	2	1	
Updated target batch-size (T_D^b)	3	3.5	3	2.7	
Target number of setups $(T_{D,k}^s)$	2	2	1	1	6

available again. The remaining unit of $D2$ is scheduled first, before the single unit of $D4$ is also assigned to machine 3. The generated schedule incurs seven setups, one more than the target value of six setups. We will use this example in the remainder to illustrate the following steps of the procedure.

The detailed procedure is shown in Fig. 6.9. For each product in order of the sequence, the procedure first initializes the remaining demand D_{pk}^r with the original product demand volume D_{pk} and the remaining number of setups S_{pk}^r with the target number of setups T_{pk}^s as calculated in Sect. 6.3.1. Then it loops until the whole demand volume of a product has been scheduled:

For each product k of family p in the given sequence $(\pi_1, \pi_2, \ldots, \pi_{K_p})$

1. Initialize data: Set remaining demand $D^r_{pk} := D_{pk}$ and remaining (target) number of setups $S^r_{pk} := T^s_{pk}$
2. Do until all units of product k have been scheduled, i.e. until $D^r_{pk} = 0$:
   a) If there is no unassigned production slot for family p in the actual sub-period, jump to next sub-period with an unassigned production slot.
   b) Calculate the number of parallel machines having an unassigned slot with a starting time in the actual sub-period (PM^u)
   c) Set number of parallel machines to schedule for product k:
      $$PM := \min\left\{ S^r_{pk}, PM^u \right\}$$
   d) Do until $PM = 0$ (load one machine per loop):
      i. Calculate batch-size for product: Set $b = \left\lceil \frac{D^r_{pk}}{PM} \right\rceil$
      ii. Determine which of the available parallel machines to load: Take any that has at least b remaining unassigned slots. If no such machine exists, take one with the highest number of remaining unassigned slots and reduce b to this figure. (If possible, take a machine that has not yet been selected for product k in the actual sub-period)
      iii. Load selected machine m with product units: Mark all used slots (in the actual and future sub-periods) as assigned
      iv. Update remaining demand: Set $D^r_{pk} := D^r_{pk} - b$
      v. Update remaining parallel machines to schedule for product k: Set $PM := PM - 1$
      vi. Update remaining target number of setups for possible later iterations of loop 2: If $S^r_{pk} > 1$ set $S^r_{pk} := S^r_{pk} - 1$

Fig. 6.9. Inner Genetic Algorithm: Deterministic loading procedure for the stage with the largest number of parallel machines (stage S)

First, it looks for a sub-period with at least one unassigned production slot and counts the number of parallel machines with unassigned slots in this sub-period (PM^u). Slots are allocated to the sub-period of their start-time. Afterwards, the algorithm calculates the number of parallel machines (PM) to load with the current product and the according batch-size b. The batch-size b is rounded to the nearest greater integer if it is a fractional number. In each of the procedure's inner loops (d), one machine is selected and loaded. Depending on b, the loading may outreach the actual and future sub-periods. If no machine exists with sufficient remaining unassigned slots, we select the machine with the highest number of remaining slots. The idea behind this is that because of the rounding up of the batch-size, the earlier batches are larger or at least equal to the later ones, implying that machines with a high number of remaining slots should be taken first. If all machines have enough unassigned slots to produce their allotted units, the procedure continues with the next product in the sequence. If at least one machine could not produce all the units, the algorithm loops to check if there are more parallel machines in the same sub-period. If there are no more machines, the next sub-period with unassigned slots is determined and the remaining units are scheduled. In both cases, we have to calculate how many additional parallel machines should be loaded for the remaining demand volume. Since the target number of setups S_{pk}^r is reduced by one with every loaded machine, the additional machines will be limited to the original target number of machines minus the number of machines that have already been loaded with the current product. S_{pk}^r is only reduced to one (and not to zero) because, if there is a remaining demand, the procedure has to load at least one more machine, even if the target number of machines has been reached.

A machine that is selected in step (d) ii. may have more available slots in the actual sub-period than are assigned in one iteration of loop (d). If another iteration of loop (d) for the same product is performed—i.e. if one more parallel machine shall be loaded for the product—this machine could be selected again. While this would reduce the number of setups (the same product would be loaded consecutively), it also implies that the target number of parallel machines would not be adhered to. We therefore only consider such machines if no other machine has unassigned slots in the actual sub-period.

Once the schedule has been generated on stage S, we roll it out to the other stages: For initialization, we create a chronological sequence of the family's production slots on each stage other than S. Again, the slots are allocated according to their start-time. Figure 6.10 gives an example for stage $S-1$ corresponding to stage S in Fig. 6.8. The numbers in the top right corner indicate the position of a slot within the sequence. We further create a chronological sequence of the product units (jobs) on stage S. In the example of Fig. 6.7, the sequence is ($C1$, $C1$, $C1$, $C1$, $C1$, $C1$, $C2$, $C3$, $C2$, $C2$, $C3$, $C2$, $C2$, $C3$, $C2$, $C4$, $C4$, $C4$, $C4$, $C4$), with the first occurrence of $C2$ coming from machine 2 and the second from machine 1, interrupted by a unit of $C3$ from machine 3. In Fig. 6.8, the sequence is ($D1$, $D1$, $D3$, $D1$, $D1$, $D3$, $D2$, $D1$, $D3$, $D2$, $D2$, $D2$, $D2$, $D2$, $D4$, $D2$).

To ensure the minimal flow time, we have to assign the jobs in the same order as they are scheduled on stage S. Thus, a straightforward approach would be to simply match all elements of the sequences in the given order, i.e. to assign the ith job scheduled on stage S to the ith slot on each of the other stages. We call this approach 'strict one-to-one matching'. Figure 6.11 gives an example. The job sequence from Fig. 6.8 is assigned to the slots

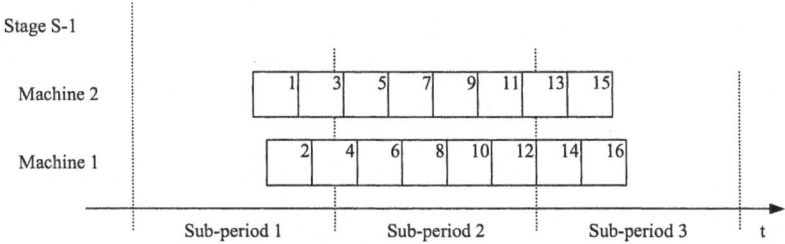

Fig. 6.10. Generating a chronological sequence of production slots on stages other than S (an example for stage $S-1$)

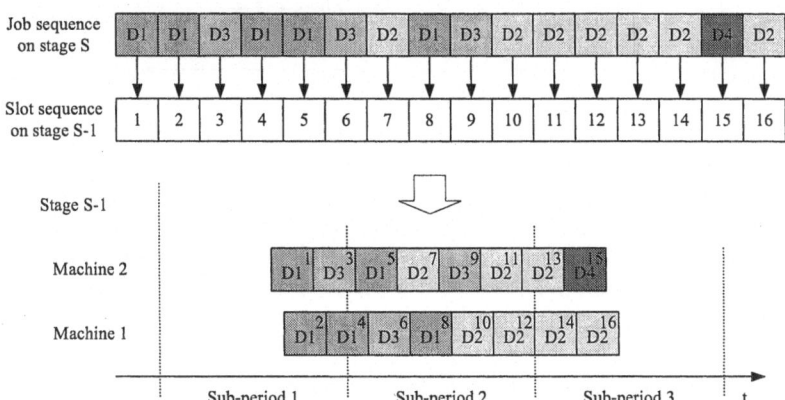

Fig. 6.11. Assigning products to slots on a stage other than S using 'strict one-to-one matching' (an example for stage $S-1$)

from Fig. 6.10. As already illustrated at the beginning of this chapter, such a procedure will in general lead to high setup costs on the stages other than S. For example, three setups are performed for product $D3$. If $D3$ was moved from slot 6 to 5 and from slot 9 to 7, all its units could be produced on a single machine with just one setup.

Hence, small changes in the sequences can save a substantial number of setups. However, every modification of the original job sequence increases the flow time. For that reason, we employ a time-based lookahead-parameter to be able to find schedules with a low number of setups and only a moderate increase of the flow time. The stages are planned independently. On each stage, the basis of the planning approach is the chronological sequence of slots. In every iteration, we take the next slot and allocate a product to it. The lookahead-parameter says the following: Suppose a slot on stage $S-1$ shall be loaded. The next unassigned job in the chronological job sequence on stage S is at position i. The start-time of the ith slot in the chronological sequence of slots on stage $S-1$ is at time t_i^{S-1}. Let l_i denote the number of slots on stage

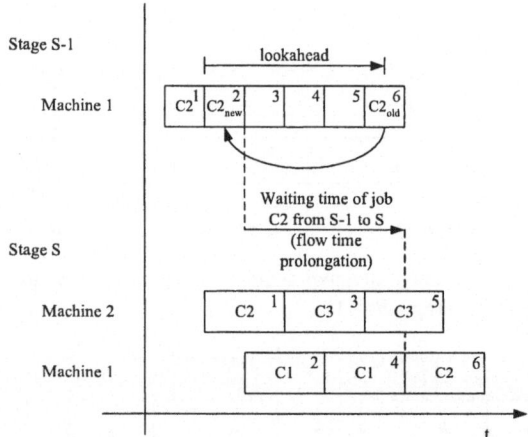

Fig. 6.12. Using a lookahead to schedule stages other than S (an example for stage $S - 1$)

$S - 1$ that have a start-time lower than or equal to $t_i^{S-1} + lookahead$ (starting with slot $i + 1$). Then, all jobs from position i to $i + l_i$ of the chronological sequence of jobs on stage S are eligible for assignment to the next available slot on stage $S - 1$. In general, the next available slot will be slot i. If so, all jobs that would be assigned to slots i to $i + l_i$ using strict one-to-one matching could be assigned to the slot at position i. However, if previous slots have been assigned using a lookahead, the current position in the slot sequence might be as high as $i + l_i$. If all slots from i to $i + l_i - 1$ receive a job using a lookahead, the slot at position $i + l_i$ must finally be loaded with job i.

Figure 6.12 gives an example covering three products ($C1$, $C2$ and $C3$). Slot 1 has been loaded with product $C2$ from the first position of the job sequence. Using strict one-to-one matching, product $C2$—from the sixth position on stage S—would be loaded on the sixth slot on stage $S - 1$. With the given lookahead, l_2 is 4 and the job can be assigned to slot 2—saving a setup because the machine is already set up for product $C2$. In the next step, slot 3 could be loaded with all jobs on position 2 to $2 + l_2 = 2 + 4 = 6$ in the job sequence. Obviously, the job at position 6 is unavailable as it has already been scheduled. Notice that—while in this example these are all jobs—the assignable jobs are counted from position 2—the first available job of the job sequence—and not from position 3—the current slot that has to be loaded. This ensures that the skipped job at position 2 is at least assigned to slot $2 + l_2$. Figure 6.12 further shows that pulling job $C2$ forward from position 6 to 2 implies a waiting time between stages $S - 1$ and S.

A job is pulled forward whenever the specific machine is set up for the associated product. If such a job is available within the lookahead, the first of these jobs is loaded on the current slot. Otherwise, no job within the lookahead can be loaded without a setup and we proceed as follows: For each job within the lookahead, we count the number of machines currently being set up for the associated product. The slot is loaded with the first job among the ones representing a product that has the lowest number of parallel machines currently set up. Thus, we favor products that are not set up on other machines. This is done because jobs representing other products may be loaded on machines that are already set up accordingly.

A job can be pulled forward by, at most, *lookahead* time units. Hence, if no other effect emerges, a job's flow time can, at most, be prolonged by *lookahead* time units. An effect further prolonging the flow time is depicted in Fig. 6.13. It shows another example for product family C. The upper panel presents the assignment using strict one-to-one matching. In the lower panel,

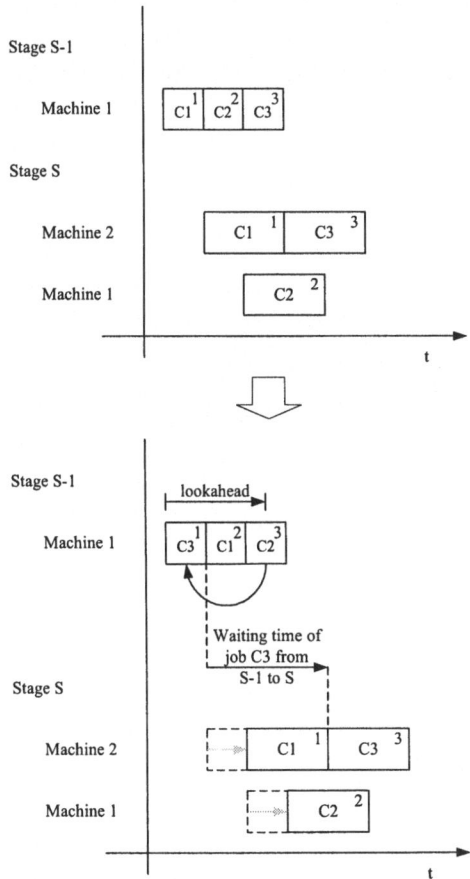

Fig. 6.13. Flow time prolongation longer than lookahead

a lookahead is used and product $C3$ is assigned to the first slot on stage $S-1$. The altered sequence of jobs on stage $S-1$ (and the unchanged sequence on stage S) implies that machine 2 on stage S must wait for stage $S-1$ to finish product $C1$ before it can produce $C1$ itself. Since on stage S, $C3$ is scheduled to be produced after $C1$ on machine 2, it also has to wait, increasing the flow time by a longer value than the given lookahead-parameter. The slots of the product family plan are postponed.

Figure 6.14 shows the above example for product family D solved using a lookahead. The lookahead-parameter is fixed for all machines and stages. However, the number of slots that can be looked ahead may vary: Slots on machine 2 are allowed to look ahead three slots ($l_i = 3$ for $i = 1, 3, 5, 7, 9, 11, 13$). For example, slot 3 can be assigned with jobs at position 3, 4, 5 or 6 in the job sequence, because slot 4 (on machine 1), slot 5 (on machine 2) and slot 6 (on machine 1) have a start-time that falls within the lookahead of slot 3. Slots on machine 1 are only allowed to look ahead two slots ($l_i = 2$ for $i = 2, 4, 6, 8, 10, 12, 14$), as the start-time of the third slot on machine 2 is just outside the lookahead. At the end of the time window, the l_i-values decrease as there are no more slots to look ahead to ($l_{15} = 1$ and $l_{16} = 0$). All possible job-to-slot assignments (using the job sequence as shown in Fig. 6.11) are listed in the second panel of Fig. 6.14. Except for the first and the last jobs, every job with an even number can be loaded on five slots (its own position as well as two slots before and after) while every job with an uneven position can be loaded on seven slots (its own position as well as three slots before and after). Arrows from later jobs to earlier slots represent assignments with a lookahead. Arrows in the other direction have to be used when a job has been skipped due to a previous assignment using a lookahead. The third panel shows the selected assignments. Job $D1$ is assigned to slot 1 on machine 2. Slot 2 is loaded with job $D3$ from position 3 because the alternative jobs on position 2 and 4 in the job sequence are both representing product $D1$, which has already been loaded on machine 2. Product $D3$ is not yet set up on any machine, and thus selected. Slot 3 can be loaded with the jobs at positions 2 and 4 because l_2 is 2 and the job at position 3 has already been loaded. Machine 2 is set up for product $D1$. Hence, we select the job at position 2 (product $D1$) as it is the first one representing this product. All jobs before position 4 are scheduled. Thus, we have all jobs from position 4 to $4 + l_4 = 4 + 2 = 6$ available for slot 4. We select $D3$ from position 6 because machine 1 is set up for it. Slot 5 is loaded with product $D1$ from position 4. For slot 6, we have $l_5 = 3$ (because the job at position 5 is the first unassigned one) and the jobs at position 5, 7 and 8 available. None of these jobs represent product $D3$, the setup state of machine 1. We select job $D2$ from position 7 as no machine is set up for it yet. Slot 7 is loaded with job $D1$ from position 5.

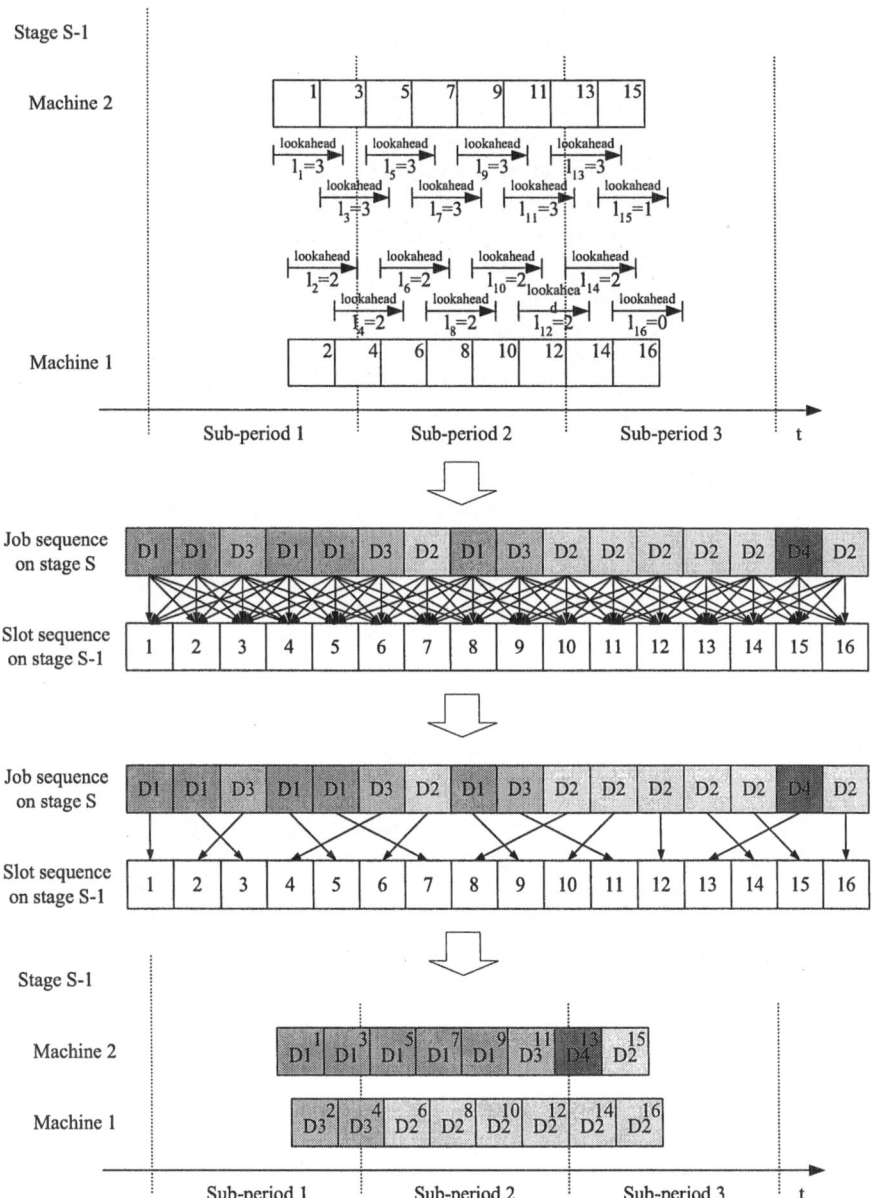

Fig. 6.14. Assigning products to slots on a stage other than S using a lookahead (an example for stage $S - 1$)

Slot 8 is loaded with product $D2$ from position 10 because machine 1 is now set up for $D2$. Slot 9 receives product $D1$ from position 8 and slot 10 product $D2$ from position 11. For slot 11, we have a choice between product $D3$

- Build sequence *JobSeq* of chronologically sorted jobs on stage S for actual product family
- Build sequence *SlotSeq* of chronologically sorted production slots on stage $S - 1$ for actual product family
- Calculate l_i, the number of slots that may be looked ahead to for all slots on stage $S - 1$ (with i representing their position in *SlotSeq*)
- Initialize *JobSeqCounter*: Set *JobSeqCounter* := 1
- For all slots o in *SlotSeq* (let m be the machine of slot o and k_m the product machine m is set up for):
  /* Select a job j to assign to slot o */

  1. For all jobs j in *JobSeq* from position *JobSeqCounter* to *JobSeqCounter* + $l_{JobSeqCounter}$
     - If job j is the first yet unassigned job representing product k_m: select job j for assignment to slot o
     - Else, if no such job exists: Select first job j such that its product is currently set up on the fewest number of parallel machines among all products in *JobSeq* from position *JobSeqCounter* to *JobSeqCounter* + $l_{JobSeqCounter}$

  /* Assign job j to slot o */

  2. Let k be the product of job j
  3. If machine m has to be set up for product k (i.e. $k \neq k_m$)
     - Set up machine m for product k
     - Increase counter of machines being set up for product k
  4. If slot o is last slot on machine m for actual product family: Decrease counter of machines being set up for product k
  5. If job was taken without lookahead (i.e. if job j is at position *JobSeqCounter* in *JobSeq*): Set *JobSeqCounter* to position of first yet unassigned job in *JobSeq* (if any)

Fig. 6.15. Inner Genetic Algorithm: Deterministic loading procedure for stage $S - 1$

from position 9 and product $D2$ from position 12. We choose $D3$ because $D2$ is already set up on machine 1 (and even if it were not, its position is after the one of $D3$). Slot 12 is loaded with product $D2$ from position 12, and slot 13 with product $D4$ from position 15 ($D2$ is already set up on machine 1). The remaining jobs all represent product $D2$. Thus, it is assigned to slots 14, 15 and 16. The resulting schedule for this stage is depicted in the lowest panel.

Figure 6.15 shows a flow chart of the procedure to schedule stage $S - 1$. In its essence, it selects a job j of stage S for each production slot o on stage $S - 1$. Job j is assigned to slot o, some parameters and counters are updated and the next slot is considered.

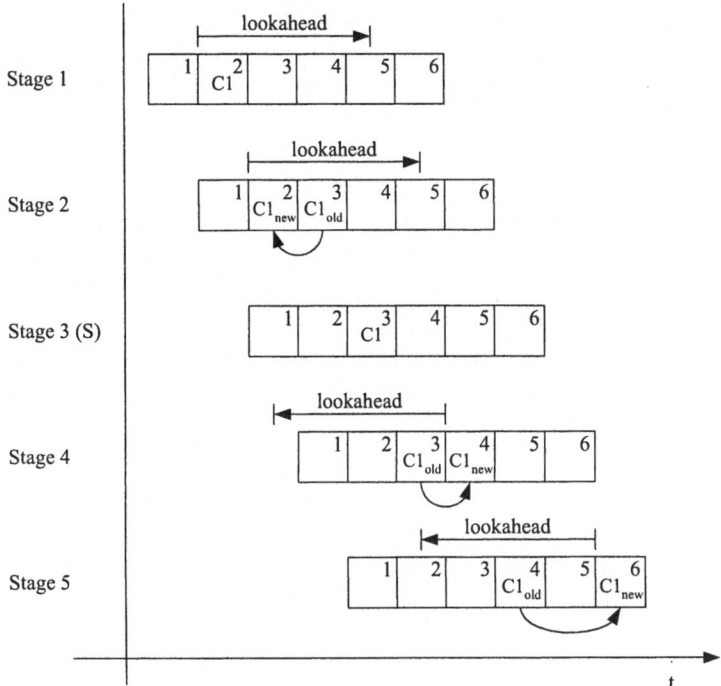

Fig. 6.16. Generating a schedule on consecutive stages

When the schedule for stage $S - 1$ has been generated, the one for stage $S - 2$ can be calculated in the same way: The jobs on stage $S - 1$ are sorted in a chronological sequence and are assigned to the chronologically sorted slots on stage $S - 2$, again using the lookahead. Hence, S is replaced by $S - 1$ and $S - 1$ by $S - 2$, and the procedure shown in Fig. 6.15 is invoked again. This method is continued until a schedule for stage 1 has been created. Stages after S are planned in an analogous manner. However, planning is done from the end to the beginning of a period and the lookahead is used to postpone jobs rather than pulling them forward. Hence, the job and slot sequences are generated in an anti-chronological order. Based on the plan for stage S, a schedule for the subsequent stage $S + 1$ is calculated and the procedure is repeated until a schedule for the last stage L has been generated. Figure 6.16 shows a condensed example for five stages with $S = 3$ being the middle stage. On stage 2, job $C1$ is pulled forward from slot 3 to 2. It remains in slot 2 on stage 1. On stage 4, it is postponed to slot 4, and on stage 5, it is assigned to slot 6. Even though the lookahead is backward-oriented for stages after S, it still starts at the beginning of a production slot. This is because all slots are allocated to their respective start-time.

It has been shown that a lookahead might postpone production slots of the product family plan. This has an impact on other product families. Their production slots have to be postponed as well, if a product family is still produced at a time when another one should be loaded according to the original product family plan. In other words, a machine may not start producing a job before the previous one is finished. Further, an operation on a previous stage must be finished before a job is started on a subsequent stage. Such a coordinated schedule on individual product level can be created using an analogous procedure to the one generating feasible machine/time slots on product family level in Phase II: The product family plan for the first stage can be used without adjustments. It represents a feasible schedule on product family level, so no two jobs overlap. There is also no earlier stage that may postpone a job's start-time. Thus, the product-to-slot assignments can directly be implemented into the product family plan. Based on the schedule for stage 1, a coordinated schedule for stage 2 can be generated using the procedure described in Sect. 5.2.2. The procedure is repeated for the other stages until a schedule for all stages has been calculated. Figure 6.17 shows the algorithm. The main difference to Phase II can be found in step 2, where a match of two

For all stages l from 1 to $L - 1$

/* Schedule stage $l + 1$ */

- Build sequence Seq^l of chronologically sorted jobs and their associated machines for all product families on stage l
- Build sequence Seq^{l+1} of chronologically sorted jobs and their associated machines for all product families on stage $l + 1$
- For all jobs j of Seq^{l+1} (let p be the product family and k the product of job j. Let m be the associated machine and t_m the time machine m becomes idle):
  1. If machine m is not set up for product family p
     - Set up machine m for product family p, starting at time t_m
     - Update the time machine m becomes idle:
       Set $t_m := t_m + t^s_{l+1, p}$
  2. Locate the first yet unused corresponding job (same product family p and product k) in Seq^l. Let t be the time it is finished on stage l
  3. Fix the start-time of job j on machine m to the minimum of t and t_m.
  4. Update the time machine m becomes idle: Set $t_m := t_m + t^u_{l+1, p}$

Fig. 6.17. Creating a coordinated schedule for all stages

operations belonging to the same job is based not only on the product family, but also on the individual product.

Using this procedure, the sequence of jobs on each machine as determined by the nested Genetic Algorithms remains unchanged, but the slots of the product family plan may be postponed. However, the product-to-slot assignment procedure can be executed with the preliminary product family plan before any postponements. This allows the product-to-slot assignment to consider the product families separately.

Figure 6.18 shows two examples. It illustrates the final schedules for stages $S - 1$ and S together. The upper panel depicts the schedule created using strict one-to-one matching (see Fig. 6.11), the lower panel the one created using a lookahead as shown in Fig. 6.14. Using a lookahead, the number of setups on stage $S - 1$ can be reduced from 11 to six. However, some jobs have a waiting time between stages. The waiting time of job $D4$ is plotted exemplarily. Most jobs are also finished later than projected in the product family plan. However, the first two jobs of product $D3$ on machine 3 can start earlier than scheduled in the product family plan—under the assumption that this does not intervene with other product families.

6.3.2.3 Evaluating a Solution

To evaluate a schedule, we calculate the intra-family setup costs on all stages. The lower the intra-family setup costs, the better the evaluation. Let S_l be the number of setups on stage l ($l = 1, 2, \ldots, L$). With c_{pl}^{si} as the setup costs for an intra-family setup of the current family p on stage l, the evaluation (fitness) of the actual schedule is calculated by $1/\sum_{l \in \{1 \ldots L\}} c_{pl}^{si} S_l$. Thus, the evaluation of an individual of the inner Genetic Algorithm—i.e. a sequence of products—is inversely proportional to the total intra-family setup costs incurred by its associated schedule.

6.3.2.4 Initial Population

The inner Genetic Algorithm's individuals are product sequences. A number of sequences is created initially.

When solving packing problems, it is generally a reasonable idea to start with large objects. The pallet is still relatively empty at the beginning and there is enough space for these objects. At the end, smaller objects can be placed into remaining empty spaces when there is not much space left. We incorporate this notion by sorting the products k of the actual family p according to their target number of setups on stage S. That is, the products are sequenced by non-increasing values of T_{pk}^s. The higher the target number of setups, the larger the 'object' that has to be placed on the pallet.

Schedule using strict one-to-one matching (11 setups on stage S-1)

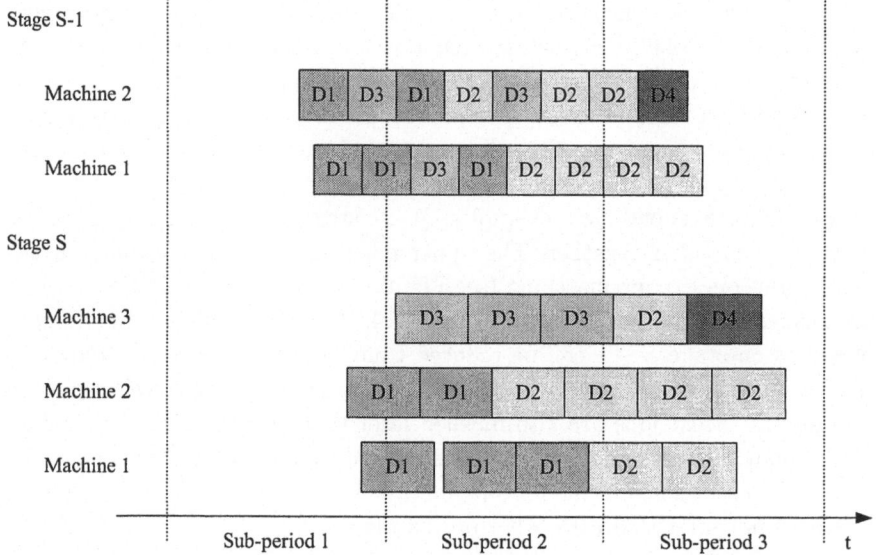

Schedule using lookahead (6 setups on stage S-1)

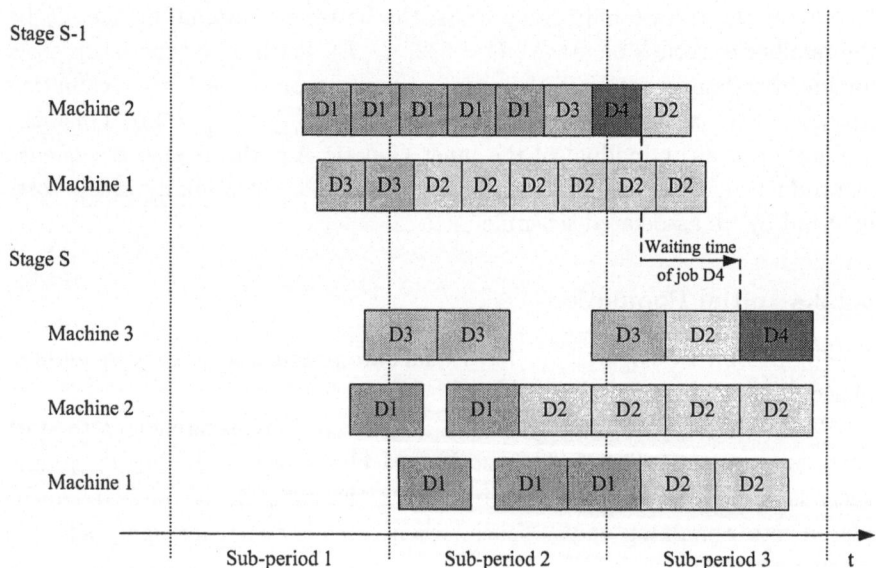

Fig. 6.18. Resulting schedules for stages $S-1$ and S, using strict one-to-one matching compared with a lookahead

The remaining sequences are created randomly. All sequences are evaluated with the deterministic loading procedure described above.

6.3.2.5 Reproduction Operators

The crossover-operator selects two product sequences at random. The selection probability is proportional to the evaluation of a sequence relative to the other sequences in the current population. The higher the evaluation, the higher the selection probability.

Let K_p be the number of products of the current family p. Thus, each sequence consists of K_p products. Two parameters, CP (copy position) and CL (copy length), are drawn at random using a uniform distribution between 1 and K_p. The first entries of the offspring sequence are created by copying CL entries from the first parent sequence, starting at position CP. The remaining entries are filled up in the order of the second parent sequence. If $CP + CL > K_p$, only the products from position CP to K_p are copied from the first parent.

Figure 6.19 shows the crossover-operator in detail. An example is given in Fig. 6.20. Starting at position $CP = 2$, $CL = 3$ products (6, 2 and 7) are taken from the father. The remaining products 3, 8, 4, 5 and 1 are filled up in the sequence of the mother.

The mutation-operator exchanges the positions of two products within a single sequence. The positions are independently drawn at random from a uniform distribution between 1 and K_p.

6.3.2.6 Creating a New Generation and Stopping Criterion

One new product sequence is generated in each iteration of the inner Genetic Algorithm. For that purpose, two individuals are selected for a crossover as described above. With a certain probability, the resulting offspring is immediately transformed by a mutation.

The new sequence is evaluated using the deterministic loading procedure. If it is better than the worst sequence of the current population, it replaces the worst one. If the new sequence is worse than the worst sequence of the current population, the new sequence is discarded. In any of the two possibilities, a new iteration is performed until no improvement could be made within a certain number of iterations or the maximum number of iterations has been reached.

Figure 6.21 shows a flowchart of the inner Genetic Algorithm.

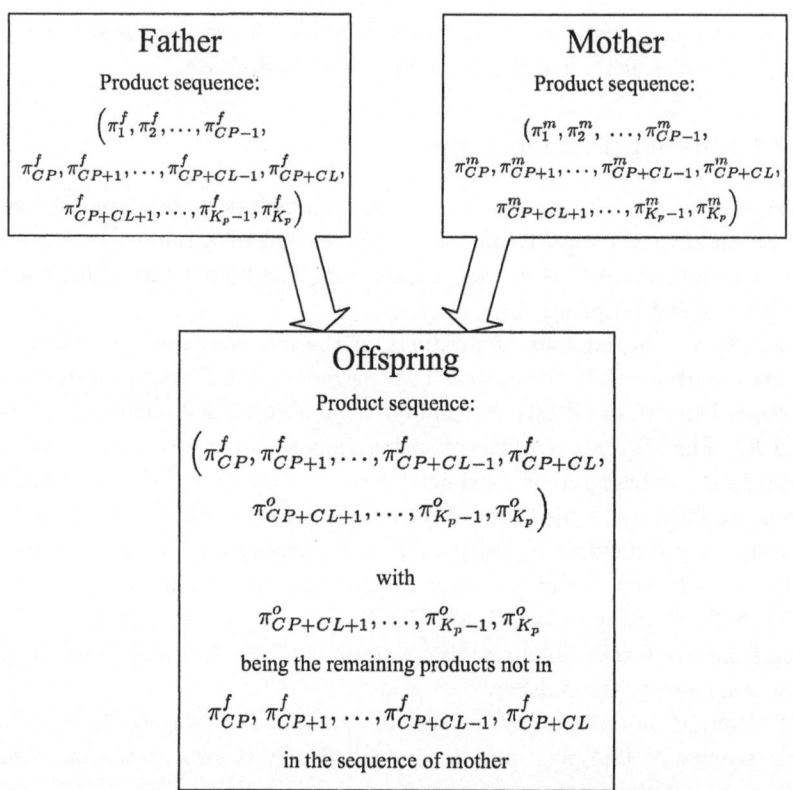

Fig. 6.19. Crossover-operator of the inner Genetic Algorithm

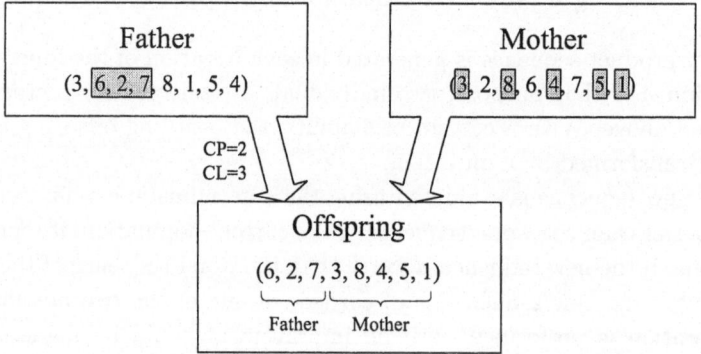

Fig. 6.20. Example of the crossover-operator of the inner Genetic Algorithm

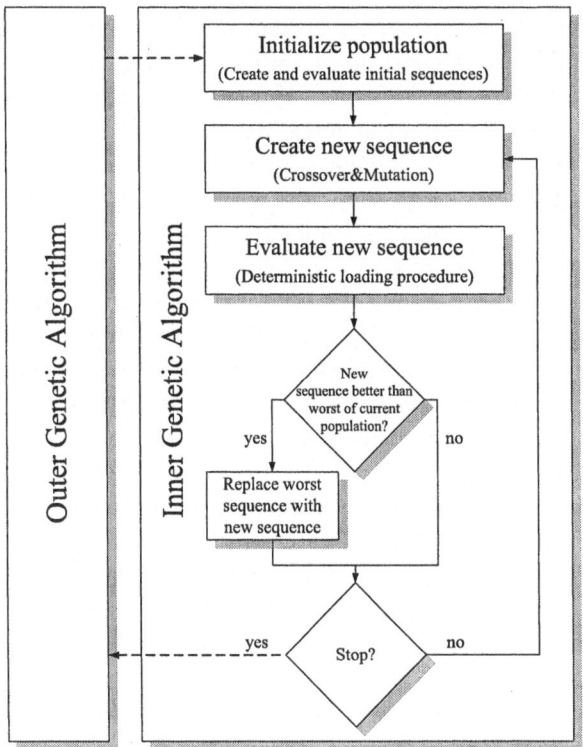

Fig. 6.21. Flowchart of the inner Genetic Algorithm

6.3.3 The Outer Genetic Algorithm: Calculating New Target Values

The inner Genetic Algorithm is coordinated by an outer Genetic Algorithm. The outer Genetic Algorithm is used to select a target number of setups (T_p^s) for each product family p from the interval $\sigma := \left[S_p^{LB}; S_p^{UB}\right]$. This figure is used to generate a target number of setups for each product within the family, and thus to determine the size of the objects for the inner Genetic Algorithm.

Each individual of the outer Genetic Algorithm's population consists of a single element, a target number of setups for the actual product family p. The initial population is created by uniformly taking PS (population size) values $s_1, s_2, \ldots, s_{PS}$ from σ with

$$s_i := S_p^{LB} + (i - 1)\frac{S_p^{UB} - S_p^{LB}}{PS - 1} , \tag{6.4}$$

where $i = 1, 2, \ldots, PS$. It is important to notice that both S_p^{LB} and S_p^{UB}, the borders of the solution space, are members of the initial population. The

solutions are evaluated with the schedule-constructing method described in
Sects. 6.3.1 and 6.3.2: In each iteration of the outer Genetic Algorithm, a
complete run of the inner Genetic Algorithm is performed. To derive the
evaluation of an individual of the outer Genetic Algorithm (coded as a target
number of setups), we take the best schedule and its evaluation according to
the inner Genetic Algorithm. Thus, the evaluation is inversely proportional
to the best schedule's total intra-family setup costs.

The crossover-operator works as follows (see Fig. 6.22): Two parent
solutions—i.e. two target number of setup values—are chosen randomly with
a probability proportional to their evaluation. An offspring is generated by
calculating the average of the two values. If the average is not integer, the
closest integer number to the result is chosen. The new target number of setups is in turn evaluated by calculating the total intra-family setup costs of the
associated schedule using the inner Genetic Algorithm. If the new solution is
better than the worst solution in the current population except for the values
S_p^{LB} and S_p^{UB}, the worst solution is replaced by the new solution. The outer
Genetic Algorithm performs another iteration until no improvement could be
made within a certain number of iterations or the maximum number of iterations has been reached. Excluding the values S_p^{LB} and S_p^{UB} ensures that
they will always remain in the population and the complete interval σ can be
reached using the crossover-operator only. For that reason, a diversification
strategy with a mutation-operator is not needed.

We employ a Genetic Algorithm in contrast to a (binary) search on the
solution space because the evaluation function need not be convex over σ.
That is, there might be local optima that we would like to overcome.

Two simple modifications can be used to speed up the computation time: If
an offspring target number of setups has been evaluated before, it is skipped
and the outer Genetic Algorithm proceeds with the next iteration. If σ is
sufficiently small—for example, if $S_p^{UB} - S_p^{LB} \leq PS$—a total enumeration

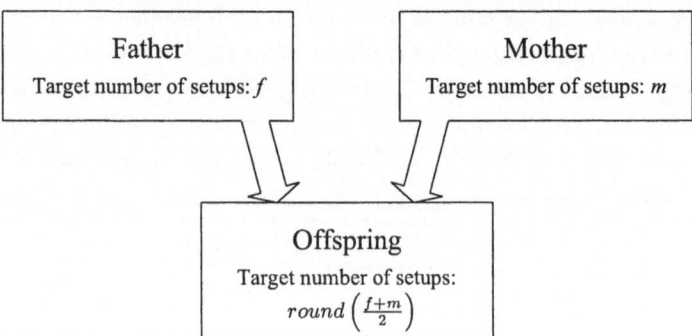

Fig. 6.22. Crossover-operator of the outer Genetic Algorithm

over all target number of setups from σ can be used instead of the outer Genetic Algorithm.

The scheduling procedure concludes with the best schedule found. The target number of setups used in the outer Genetic Algorithm as well as the product sequence of the inner Genetic Algorithm are discarded as they are only a means to an end. The scheduling procedure is invoked for each product family and the final result is the coordinated schedule for all product families.

6.4 Computational Results

In this section, we test the solution quality and computation time of the developed procedures. The Genetic Algorithms contain a number of setting-parameters that have to be optimized. Associated tests are reported on in Sect. 6.4.1. In Sect. 6.4.2, we report on computational tests for various problem characteristics: We use and expand the instances employed for the tests of Phase I and their respective solutions. In this way, these instances undergo the complete planning approach. Since they consist of several product families, we can simultaneously test the procedures developed for Phase II and Phase III.

There are two measures for the solution quality of a schedule: The number of setups incurred and the (average) relative flow time (RFT), which is flow time divided by raw process time. The relative flow time is an indicator of a job's waiting time between stages. It is always equal or greater than one, and equals one if a job does not have to wait between production stages.

All tests have been performed on an Intel Pentium IV processor with 2.2 MHz and 1 GB RAM. The procedures developed for Phases II and III have been implemented in Microsoft Visual Basic 4.

6.4.1 Parameter Optimization

We create 10 instances to optimize the various setting-parameters of the Genetic Algorithms. The instances consist of one period, a single product family, and three production stages. There are 10 machines on the first stage, 20 machines on the second stage, and 2 machines on the third stage. The third stage constitutes the bottleneck. The planning period is 8 000 time units. The instances are created in a way that for the first half of the planning period, both machines on the third stage produce the product family. After 4 000 time units, one machine finishes production and the product family is only produced on the other machine. This machine finishes at the end of the planning period (at time 8 000). We find this is a typical scenario for larger instances containing several product families.

The process times are 50, 100 and 10 time units for stage 1, 2 and 3, respectively. Since there is only one product family, setup times are irrelevant. The product family consists of 10 products. While intra-family setups have to be performed on stages 1 and 2, the machines on stage 3 do not differentiate between the products of a family. Thus, no setup has to be performed when changing from one product to another and we do not count (intra-family) setups on this stage. On stages 1 and 2, intra-family setup costs are equal.

Preliminary demand figures are drawn at random using a uniform distribution between 20 and 220. These figures are then scaled in a way that the total demand per instance adds up to 1200 units (implying that a demand volume of a product may lie outside the interval [20; 220]). 800 of these units are produced on one of the machines of stage 3, while the other machine produces 400 units.

This data leads to a minimum target number of setups of $S_p^{LB} = 10$ and a maximum target number of setups of $S_p^{UB} = 10 \cdot 20 = 200$ (the instances do not contain a demand figure less than 20, so the number of machines on stage 2 (=20) is used for the calculation). Thus, the solution space of the outer Genetic Algorithm consists of 190 individuals. The inner Genetic Algorithm may evaluate $10! = 3\,628\,800$ product sequences. We assume that the first production on a machine requires an intra-family setup. Hence, initial intra-family setup states are not considered. Cyclic production (Phase II) is not allowed.

6.4.1.1 Number of Generations

We use the following settings to test the effect of the number of generations for the Genetic Algorithms: The population size of the outer Genetic Algorithm is set to 10, the one of the inner Genetic Algorithm to 50. The mutation probability of the inner Genetic Algorithm is set to 50%. The lookahead-parameter is set to 50 time units, and the length of a sub-period is 2 000 time units, implying there are four sub-periods within the planning interval.

Figure 6.23 shows the total number of product setups (summed up for stages 1 and 2) for an increasing number of generations of the outer Genetic Algorithm, including the 10 evaluations for the initial population. For each generation, the final number of setups of the best solution found so far is displayed. Each line corresponds to one of the 10 instances, and the bold line represents the average. For an easy comparison, the figures have been scaled: For each instance, the number of setups of the first solution is set to 100%.

The graphs are typical for Genetic Algorithms: The larger the number of generations, the better the results—but with a larger number of generations, the improvements diminish. For one instance, no improvement can be made by increasing the number of generations. Its graph remains at 100%. For this instance, the first target number of setups that had been evaluated already delivered a very good solution. However, in general, one can clearly see the effectiveness of the outer Genetic Algorithm as the number of generations increases.

Figure 6.24 shows an analogous graph for the computation time (in h:mm:ss) of the outer Genetic Algorithm. Even though the graph is flattening for a high number of generations, the computation time in fact increases almost linearly with the number of calls to the inner Genetic Algorithm. However, the inner Genetic Algorithm is not called if a specific target number of setups—i.e. an individual of the outer Genetic Algorithm—has already been evaluated in an earlier generation. Obviously, the probability for this increases with the number of generations: In later generations, the outer Genetic Algorithm creates similar individuals as it tries to intensify the search in the promising regions of the solution space.

The inner Genetic Algorithm is analyzed in Fig. 6.25. For each run of the outer Genetic Algorithm, there are a number of runs of the inner Genetic Algorithm. We have chosen to display the run for the best target number of setups for each instance. The number of setups decreases very smoothly with the number of generations. A graph for the computation time is omitted because each of the inner Genetic Algorithm's generations consumes almost a constant, short amount of time.

With the computation times in mind, we set the number of generations for the outer Genetic Algorithm to 30 and the number of generations for the inner Genetic Algorithm to 4 000 for the remaining setting-parameter tests. A higher number of generations does not seem to improve the solution quality substantially. In relative terms, this means that the number of generations for the outer Genetic Algorithm is set to about 15% of the size of its solution space (30/190), and the number of generations for the inner Genetic Algorithm to about 0.1% of the size of its solution space (4 000/3 628 800).

6.4.1.2 Population Size

For this and the following sections, all parameter settings (except the ones being investigated) are as defined in Sect. 6.4.1.1. An exception is made for the mutation probability of the inner Genetic Algorithm, which is set to 60% in order to allow uniform intervals when testing this parameter.

Table 6.3 shows the results for a varying population size of the outer Genetic Algorithm. We tested population sizes of 4, 7, 10, 13, 16 and 19. The table lists the total number of product setups summed up for stages 1 and 2 (#Setups), its standard deviation (StdDev Setups), the average relative flow time (RFT) and associated standard deviation (StdDev RFT) as well as the mean computation time in h:mm:ss (Time). Figure 6.26 displays the

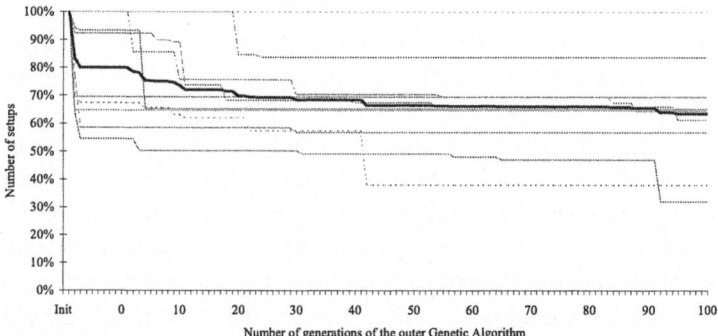

Fig. 6.23. Effect of the number of generations of the outer Genetic Algorithm on the total number of setups

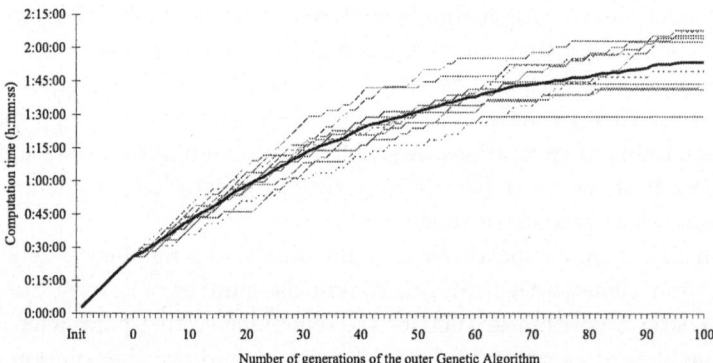

Fig. 6.24. Effect of the number of generations of the outer Genetic Algorithm on the computation time

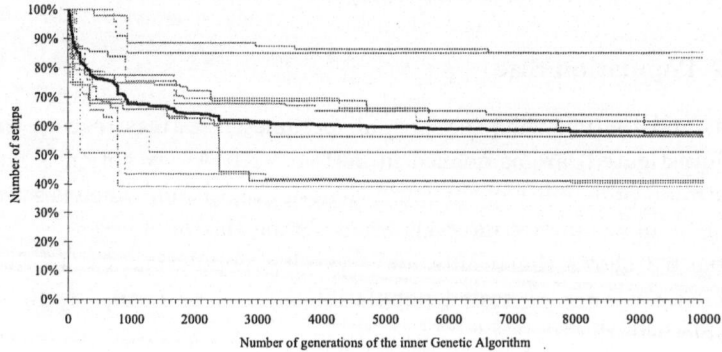

Fig. 6.25. Effect of the number of generations of the inner Genetic Algorithm on the total number of setups

total number of setups, Fig. 6.27 the computation time. As can be seen, the number of setups does not seem to change systematically with the population size, but the computation time increases the larger the population size. This is obvious because the larger the population size, the more individuals have to be evaluated initially. As can be seen from the table, the relative flow time remains relatively unaffected.

Table 6.4 and Figs. 6.28 and 6.29 show the results for a varying population size of the inner Genetic Algorithm. We tested population sizes of 50, 70, 90 and 500. The results are very similar to the outer Genetic Algorithm: The number of setups as well as the relative flow time seem not to be affected systematically, and the computation time increases. However, for the inner Genetic Algorithm, the prolongation of computation time is less strong. One can only see the effect for a population size of 500.

6.4.1.3 Mutation Probability

The inner Genetic Algorithm makes use of a mutation-operator which—with a certain probability—is invoked after a crossover. Mutation probability settings of 0%, 20%, 40%, 60%, 80% and 100% have been evaluated. The results are shown in Table 6.5 and Figs. 6.30 and 6.31. The mutation probability does not seem to have any systematical effect on the number of setups, the relative flow time or the computation time. It can therefore be set arbitrarily.

6.4.1.4 Lookahead

The Genetic Algorithms employ a lookahead-parameter to manage the trade-off between the objectives of a low number of setups and a low flow time. We tested lookahead-parameters of 0, 50, 100, 200, 400, 800 and 1600 time units. These absolute figures have to be set in relation to the process time on the first stage, which is 50 time units. Hence, the lookahead is set to 0, 1 time, 2 times, 4 times, 8 times, 16 times and 32 times the process time on stage 1. The results are listed in Table 6.6.

Figure 6.32 shows the total number of setups. As expected, it decreases substantially when the lookahead becomes longer. For a lookahead-parameter above zero, one can see that the number of setups decreases by a relatively constant amount whenever the lookahead is doubled. Figure 6.33 shows the effect on the relative flow time. As derived theoretically, the relative flow time increases with a longer lookahead. Even though the curve shows an exponential increase, one should also notice that the horizontal axis is scaled exponentially. The same has to be said about the computation time, which is depicted in Fig. 6.34. It also increases for a longer lookahead, because loop 1 in the algorithm of Fig. 6.15 will in general perform more iterations.

Table 6.3. Effect of the population size of the outer Genetic Algorithm

Population size	#Setups	StdDev Setups	RFT	StdDev RFT	Time
4	84.2	16.83	1.248	0.076	0:16:03
7	82.7	13.92	1.247	0.048	0:24:58
10	85.4	15.30	1.250	0.060	0:28:57
13	85.6	12.53	1.271	0.075	0:33:24
16	91.5	13.07	1.265	0.064	0:36:36
19	86.2	13.85	1.226	0.042	0:39:28

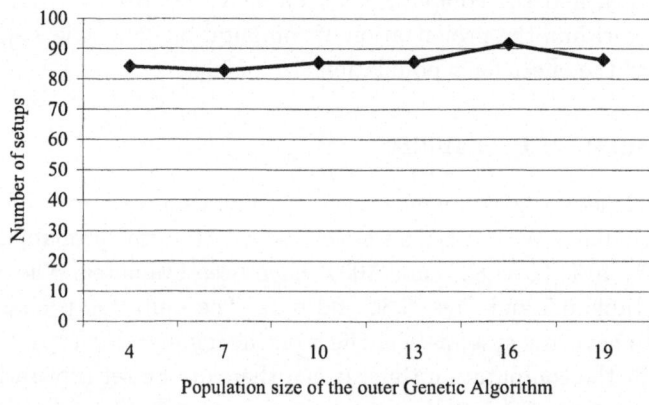

Fig. 6.26. Effect of the population size of the outer Genetic Algorithm on the total number of setups

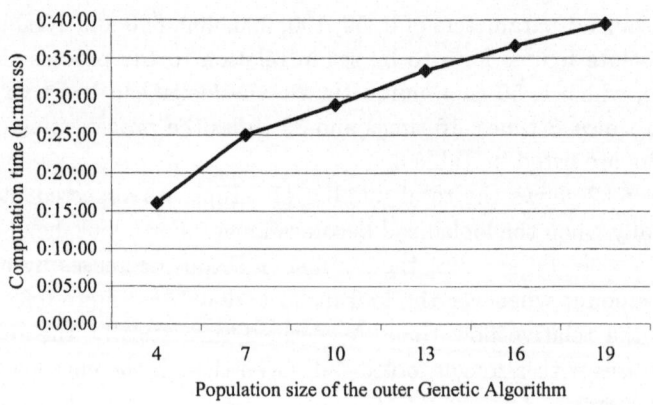

Fig. 6.27. Effect of the population size of the outer Genetic Algorithm on the computation time

Table 6.4. Effect of the population size of the inner Genetic Algorithm

Population size	#Setups	StdDev Setups	RFT	StdDev RFT	Time
10	89.9	15.88	1.257	0.046	0:28:28
30	88.0	13.58	1.247	0.049	0:28:43
50	85.4	15.30	1.250	0.060	0:28:57
70	89.3	9.55	1.263	0.071	0:29:19
90	92.2	10.69	1.284	0.072	0:28:50
500	93.2	11.97	1.257	0.055	0:33:35

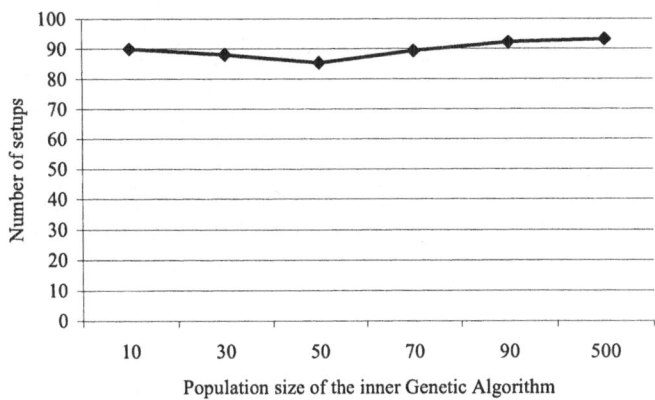

Fig. 6.28. Effect of the population size of the inner Genetic Algorithm on the total number of setups

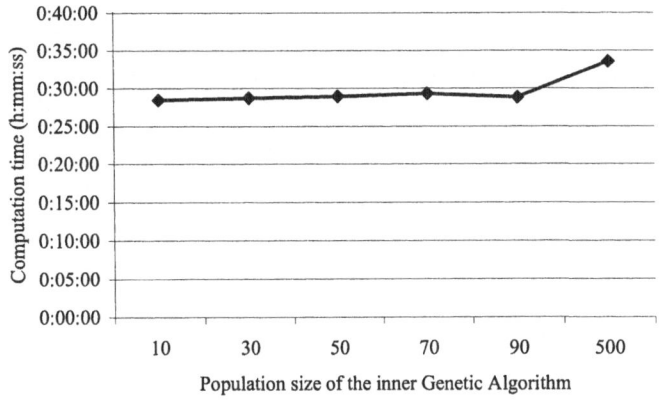

Fig. 6.29. Effect of the population size of the inner Genetic Algorithm on the computation time

Table 6.5. Effect of the mutation probability of the inner Genetic Algorithm

Mutation prob.	#Setups	StdDev Setups	RFT	StdDev RFT	Time
0%	91.4	14.09	1.258	0.049	0:28:33
20%	87.4	11.38	1.271	0.066	0:27:45
40%	90.2	13.34	1.265	0.054	0:28:04
60%	85.4	15.30	1.250	0.060	0:28:57
80%	86.6	15.70	1.265	0.074	0:28:41
100%	89.8	15.19	1.256	0.047	0:28:35

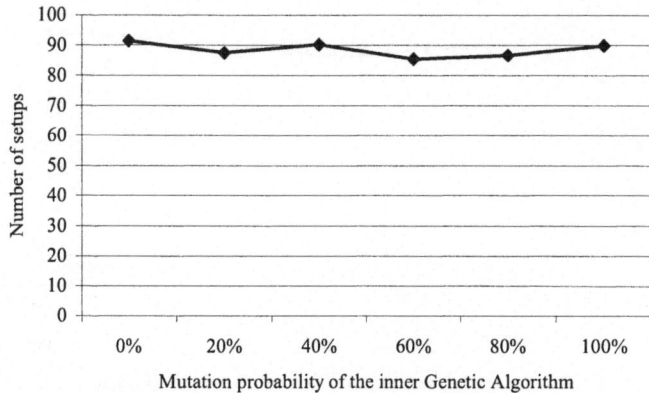

Fig. 6.30. Effect of the mutation probability of the inner Genetic Algorithm on the total number of setups

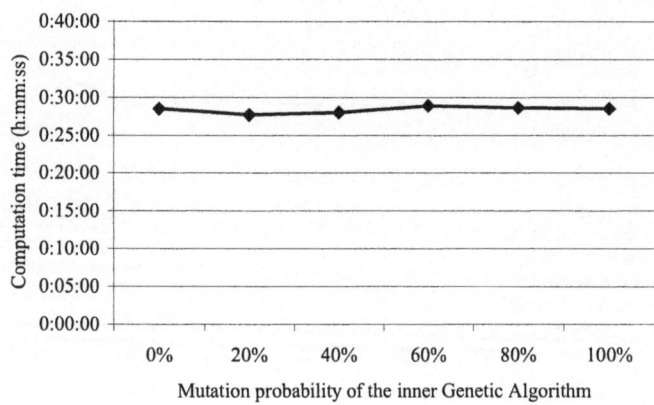

Fig. 6.31. Effect of the mutation probability of the inner Genetic Algorithm on the computation time

Table 6.6. Effect of the lookahead-parameter

Lookahead	#Setups	StdDev Setups	RFT	StdDev RFT	Time
0	284.6	19.72	1.229	0.000	0:27:42
50	85.4	15.30	1.250	0.060	0:28:57
100	79.7	10.13	1.350	0.102	0:32:08
200	68.5	8.24	1.420	0.119	0:37:32
400	62.6	7.00	1.631	0.262	0:48:13
800	56.5	4.77	2.186	0.389	1:15:31
1600	51.8	2.10	2.701	0.466	2:03:53

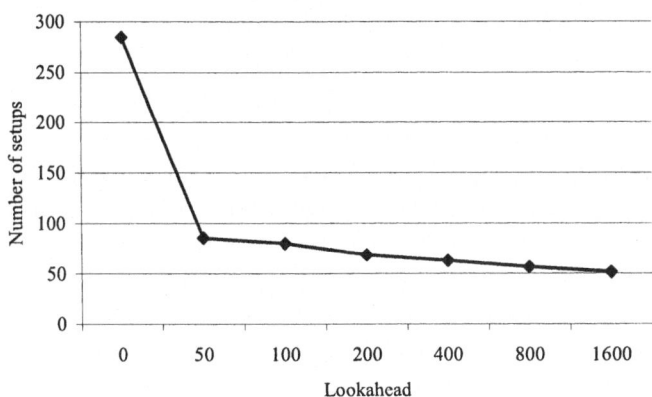

Fig. 6.32. Effect of the lookahead-parameter on the total number of setups

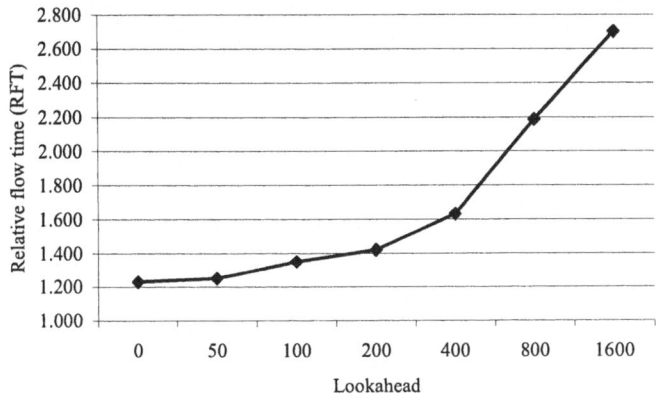

Fig. 6.33. Effect of the lookahead-parameter on the relative flow time

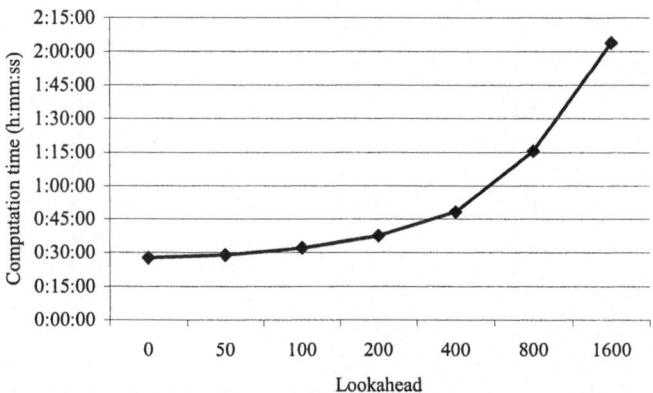

Fig. 6.34. Effect of the lookahead-parameter on the computation time

As a summary, the figures show that the lookahead seems to be a good parameter to manage the trade-off between the objectives of a low number of setups and a low flow time. Because of this, there is no 'right' or 'wrong' setting for this parameter. It has to be fixed by a human planner in order to achieve results that are congruent with the specific situation and objectives. It seems reasonable to relate the parameter to the average process time on stage 1.

Further tests have shown that, with a longer lookahead, the Genetic Algorithms become less effective. That is, the number of setups decreases only very slightly (or not at all) with an increasing number of generations. Displayed in the way as in Fig. 6.23, this would result in almost horizontal lines, for both the inner and the outer Genetic Algorithm. For the outer Genetic Algorithm, this behavior can be explained as follows: In general, the best solution—i.e. the lowest number of resulting setups—is achieved by the smallest possible target number of setups (S_p^{LB}). Obviously, this leads to very few setups on stage S. Using 'strict one-to-one matching', a high number of setups on the other stages would be the result. With a long lookahead, these setups can be avoided by local re-sequencing. Ultimately, with a lookahead close to the period length, only very few setups have to be performed on these stages as well. Hence, the outer Genetic Algorithm has no effect as the minimal number of setups is already achieved by the minimal target number of setups, which is always evaluated as the first candidate. Further, the different product sequences of the inner Genetic Algorithm result in a very similar number of setups. Thus, the inner Genetic Algorithm is also able to find good solutions at a very early stage.

6.4.1.5 Number of Sub-Periods

The final setting-parameter tested is the number of sub-periods used for both Genetic Algorithms. We tested a single, 2, 4, 8 and 16 sub-periods (thus, the length of a sub-period has been set to 8 000, 4 000, 2 000, 1 000 and 500 time units, respectively). The results are depicted in Table 6.7 as well as in Figs. 6.35 and 6.36. The algorithms seem to be relatively unaffected by the number of sub-periods. Concerning the computation time, there seems to be a slight advantage for a medium number of sub-periods.

6.4.1.6 Selecting 'Optimal' Parameter Values

The tests have shown that the number of generations for the outer and the inner Genetic Algorithm, as well as the lookahead and—to a smaller extent—the number of sub-periods, have an effect on solution quality and computation time. The other setting-parameters (population size and mutation probability) seem not to affect the algorithms systematically. Thus, the latter are fixed at a reasonable level, while the former will be determined for each instance or product family individually. All settings—and hence a summary of the results for the parameter optimization—are depicted in Table 6.8.

6.4.2 Effect of Problem Characteristics

The tests reported on in this section assess the solution quality and computation time of the algorithms for varying problem characteristics. This is done using the instances and results generated for Phase I, implying that these problems undergo the complete solution approach.

6.4.2.1 Experimental Design and Instance Generation

Analogously to the parameter optimization, we consider three production stages and the machines on stage 3 do not differentiate between products of a family. Thus, no setup has to be performed when changing from one product to another and we do not count (intra-family) setups on this stage. The third stage is also supposed to be the bottleneck stage which has been considered in Phase I of the solution approach. This configuration is derived from practical business cases.

We use the small and the large test problems from Chap. 4. The small test problems cover 5 product families, 4 periods and 4 machines on the bottleneck stage. The large test problems consist of 20 product families, 6 periods and 10 machines on the bottleneck stage. The factors product family 'demand variation', 'inventory holding costs' and product family 'demand probability'

Table 6.7. Effect of the number of sub-periods

#Sub-periods	#Setups	StdDev Setups	RFT	StdDev RFT	Time
1	94.1	20.66	1.261	0.089	0:31:16
2	88.2	12.98	1.253	0.077	0:29:05
4	85.4	15.30	1.250	0.060	0:28:57
8	93.8	11.25	1.294	0.059	0:28:41
16	82.9	15.19	1.227	0.033	0:30:37

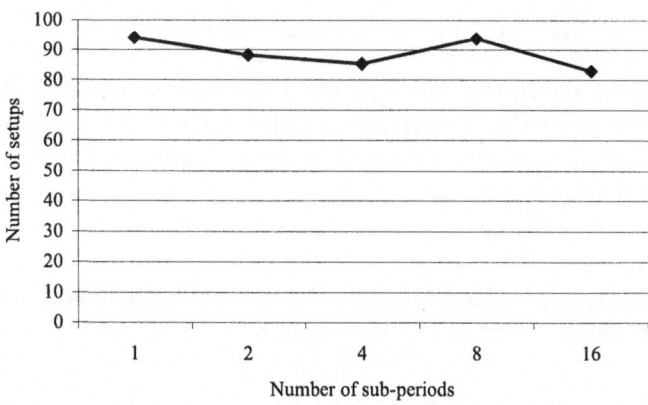

Fig. 6.35. Effect of the number of sub-periods on the total number of setups

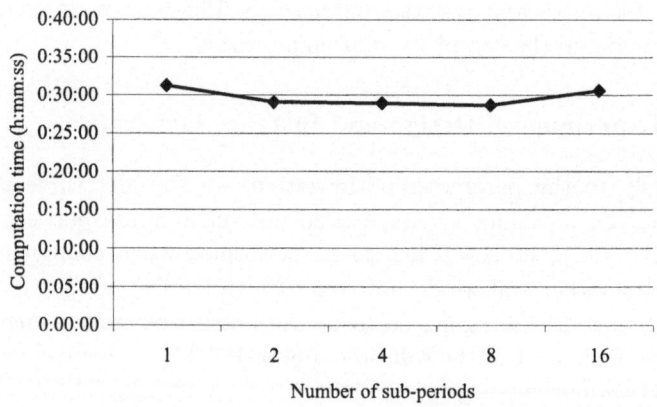

Fig. 6.36. Effect of the number of sub-periods on the computation time

Table 6.8. Results for the parameter optimization (parameters marked with an asterisk (*) will be calculated for each instance or product family individually)

Setting-parameter	Outer Genetic Algorithm	Inner Genetic Algorithm
Number of generations*	15% of the size of its solution space, minimally 100	0.1% of the size of its solution space, minimally 10
Population size	7	50
Mutation probability	not applicable	60%
Lookahead*	Once the average process time on stage 1	
Number of sub-periods*	4 (length of a sub-period is set to a quarter of a period)	

from Phase I are not relevant for Phases II and III. Thus, the instances having a high demand variance, high inventory holding costs or a low demand probability are deleted from the sample. For the same reason, we also delete all instances showing a positive demand trend, because Phases II and III cover just a single period. Further, we only consider the first of three instances for each factor combination of the large test problems.

The algorithms in Phases II and III are based on a solution for Phase I. We use the solutions generated by procedure LSP-WSCF as a basis. Considering only the solved instances, this leads to a sample of 76 instances for the small test problems (four instances have not been solved in Phase I), and eight instances for the large test problems. However, all periods of the original problems are treated as a separate problem for Phases II and III. For the small problem sets, which contain four periods each, this leads to 304 problems. For the large problem sets with six periods each, it results in 48 problems. As the original instances cover only a single production stage, additional data has to be determined. For each of the problems, we create $2 \times 2 \times 2 = 8$ instances based on three additional factors that are relevant for Phases II and III:

1. **Products per family:** In the low number of products per family case, we have 4 products per family. In the high number of products per family case, there are 10 products per family.
2. **Number of machines/machine configuration:** We consider two scenarios, both representative for practical configurations. In the first sce-

Table 6.9. Number of machines for the small and the large problem sets

Problem sets	Scenario	#Machines		
		Stage 1	Stage 2	Stage 3
small	1	4	6	4
	2	16	32	4
large	1	10	15	10
	2	40	80	10

nario, the number of machines on stage 1 is set equal to the number of machines on the bottleneck stage (stage 3). The number of machines on stage 2 is set to 1.5 times the number of machines on the bottleneck stage. In the second scenario, the number of machines on stage 1 is set to four times the number of machines on the bottleneck stage, and the number of machines on stage 2 to eight times the number of machines on the bottleneck stage. The resulting number of machines for the small and the large problem sets are depicted in Table 6.9.

We call scenario 1 the 'low number of machines' case and scenario 2 the 'high number of machines' case.

3. **Product demand pattern:** Analogously to the product families, we differentiate between a demand that is equally distributed among the products of a family and a 'different sizes' scenario (the 'positive trend' case makes no sense in a single period environment). In the equally distributed case, the demand of product k of family p in period t is set to

$$D_{pkt} := \text{round}\left(\frac{\sum_{m \in \bar{M}} x_{ptm}}{K_p}\right)$$
$$\forall\, p \in \bar{P},\ t \in \bar{T},\ k \in \{1, 2, \ldots, K_p - 1\}, \quad (6.5)$$

with x_{ptm} the production volume of product family p on machine m on the bottleneck stage in period t (a result of Phase I) and K_p the number of products in family p. In the 'different sizes' case, the demand is set to

$$D_{pkt} := \text{round}\left(\frac{\sum_{m \in \bar{M}} x_{ptm}}{K_p} \cdot \frac{2k}{K_p + 1}\right)$$
$$\forall\, p \in \bar{P},\ t \in \bar{T},\ k \in \{1, 2, \ldots, K_p - 1\}. \quad (6.6)$$

Further explanations can be found in Sect. 4.5.1.1. In order to avoid potentially negative demand volumes, the demand of the last product of each family is set to the remaining units:

$$D_{pK_pt} := \sum_{m \in \bar{M}} x_{ptm} - \sum_{k=1}^{K_p-1} D_{pkt} \quad \forall p \in \bar{P}, t \in \bar{T} \tag{6.7}$$

We keep three factors from the experimental design for Phase I:

4. **Product family demand pattern:** The 'equal' and the 'different sizes' product family demand patterns are retained.

5. **Capacity utilization:** We vary the capacity utilization by different process times. For each product family p and stage l (except the bottleneck stage B), preliminary process times ($\tilde{t}_{lp}^u$) are drawn at random from a uniform distribution between 0.8 and 1.2. The average processing time over all machines and periods on stage l is calculated by

$$APT_l^{MT} := \sum_{\substack{p \in \bar{P} \\ t \in \bar{T}}} \left(\tilde{t}_{lp}^u \cdot D_{pt} \right) / (M_l \cdot T) \quad \forall l \in \{1, 2, \ldots, L\} \setminus B, \tag{6.8}$$

with $D_{pt} := \sum_{k=1}^{K_p} D_{pkt}$ the demand of family p in period t and M_l the number of machines on stage l. The final process times are calculated by

$$t_{lp}^u := \tilde{t}_{lp}^u \cdot \frac{C}{APT_l^{MT}} \cdot CUF \cdot 0.65 \quad \forall p \in \bar{P}, l \in \{1, 2, \ldots, L\} \setminus B, \tag{6.9}$$

with C the period-capacity and CUF the capacity utilization factor on the bottleneck stage (0.7 in the low capacity utilization case and 0.85 in the high capacity utilization case).

Verbally, this means that the capacity utilization (caused by processing only) is set to 65% of the capacity utilization on the bottleneck stage.

6. **Product family setup time:** Setup times are only considered between product families. Intra-family setups do not consume machine capacity. Analogously to the bottleneck stage, product family setup times are relative to processing times. We calculate the average processing time over all periods caused by an average product on stage l by

$$APT_l^{PT} := \sum_{\substack{p \in \bar{P} \\ t \in \bar{T}}} \left(t_{lp}^u \cdot D_{pt} \right) / (P \cdot T) \quad \forall l \in \{1, 2, \ldots, L\} \setminus B. \tag{6.10}$$

Setup times are generated with the formula $t_{lp}^s := APT_l^{PT} \cdot U(1; 3) \cdot S$ for all products $p \in \bar{P}$ and stages $l \in \{1, 2, \ldots, L\} \setminus B$. $U(1; 3)$ indicates a uniform distribution between 1 and 3. S is held at two levels of 0.01 (low setup times) and 0.05 (high setup times). For the low setup time case, this leads to setup times which are between 1% and 3% of the per period processing time of an average product. For the high setup time case, the setup times are between 5% and 15% of this figure.

Together, this makes $2 \times 2 \times 2 \times 2 \times 2 \times 2 = 64$ factor combinations and leads to 2432 instances for the small problem sets and to 384 instances for the large problem sets. Table 6.10 summarizes the factorial design.

The remaining data is set as follows: For each instance representing the first period of the original lot-sizing and scheduling problem, initial product family setup states on stages 1 and 2 are selected randomly (on stage 3, initial setup states have already been determined). The product family setup state is carried over between periods. For the computational tests, we assume that product family setups also count as intra-family setups because the machine has to be set up not only for the family, but also for the respective product. Further, we assume that the setup state on product level cannot be carried over between periods. Thus, an intra-family setup is counted for each machine that starts production in a period. However, it has been made sure that production in a later period does not begin before the respective machine has completed the production volume of a previous period.

The setting-parameters are fixed as described in the previous section: The lookahead and the number of sub-periods are determined for each instance, the number of generations for each product family individually. Intra-family setup costs are identical for stages 1 and 2 and all product families. Cyclic production (Phase II) is not allowed.

Table 6.10. Experimental design for the tests of Phases II and III

Products per family (#Products)	low number of products per family: 4 high number of products per family: 10
Number of machines/ machine configuration (#Machines)	low number of machines: 4-6-4, 10-15-10 high number of machines: 16-32-4, 40-80-10
Product demand pattern (ProdDemPat)	equal different sizes
Product family demand pattern (FamDemPat)	equal different sizes
Capacity utilization (CapUtil)	low capacity utilization: $APT_l^{MT} / (0.65 \cdot 0.7)$ high capacity utilization: $APT_l^{MT} / (0.65 \cdot 0.85)$
Product family setup time (SetupTime)	low setup times: $APT_l^{PT} \cdot U(1;3) \cdot 0.01$ high setup times: $APT_l^{PT} \cdot U(1;3) \cdot 0.05$

6.4.2.2 Small Test Problems

The small test problems contain 5 product families. For the instances with a low number of products per family, this leads to 20 products in total. For the instances with a high number of products per family, 50 products have to be scheduled. The number of machines on each of the three stages is depicted in Table 6.9. On average, a total of 500 jobs have to be scheduled.

Table 6.11 displays the results. It shows the average number of setups (#Setups) and the associated standard deviation (StdDev), the average relative flow time (RFT) and its standard deviation (StdDev) as well as the average computation time in h:mm:ss (Time). A single asterisk (*) denotes a significant difference at a 5% error level, a double asterisk (**) at an 0.5% error level. Tests have been performed using unpaired t-tests.

The number of setups is significantly influenced by the number of products per family, the number of machines, and the product demand pattern: It is obvious that more setups have to be performed when a higher number of products have to be scheduled. However, the increase is subproportional, because the total demand (500 jobs on average) remains unchanged and thus, each product has a lower demand figure. The effect of the number of machines can be explained as follows: For a high number of machines, the instance generation has led to longer process times than for a low number of machines and thus, more machines (and setups) are needed to produce the demand volume. The effect of the product demand pattern is relatively small (139.9 setups on average in the equally distributed case and 120.6 in the 'different sizes' case). There is also a slight difference in the means for the product family demand pattern. However, this effect is not statistically significant. The other two factors—capacity utilization and product family setup times—seem not to have any effect on the number of intra-family setups.

The relative flow time (RFT) is significantly influenced by the number of machines, the product family demand pattern, the capacity utilization, and the product family setup times. The strongest effect is exercised by the setup times. This is because long setup times cause long waiting times when a machine has to be set up for a new product family. As seen in Sect. 6.4.1.4, the relative flow time may be decreased by a longer lookahead.

The computation time is affected only by the number of products per family and the number of machines. These factors have a direct influence on the problem size and it seems obvious that larger problems need longer computation times. Increasing the number of products from 4 to 10 implies that the solution space of the inner Genetic Algorithm is increased from $4! = 24$ to $10! = 3\,628\,800$ for each of the families, which explains the large prolongation of computation times. The other factors seem to have no influence on the computation time, which is a remarkable result, as it implies that the run-times

Table 6.11. Factorial design results for small test problems

	#Setups	StdDev	RFT	StdDev	Time
#Products					
low	95.8**	42.84	2.867	2.075	0:00:22**
high	164.6	64.37	2.943	2.248	0:07:03
#Machines					
low	82.9**	32.44	2.588**	1.435	0:02:10**
high	177.6	52.99	3.222	2.665	0:05:15
ProdDemPat					
equal	139.9**	70.49	2.905	2.247	0:03:46
diff sizes	120.6	56.51	2.906	2.077	0:03:39
FamDemPat					
equal	132.5	66.18	3.047**	2.304	0:03:47
diff sizes	127.8	62.81	2.756	1.994	0:03:38
CapUtil					
low	130.6	66.90	2.493**	1.452	0:03:41
high	129.9	61.96	3.363	2.672	0:03:44
SetupTime					
low	130.8	65.69	1.652**	0.593	0:03:44
high	129.6	63.37	4.298	2.409	0:03:41
Total	130.2	64.59	2.905	2.163	0:03:43

of the procedures do not depend on certain problem characteristics such as capacity utilization.

6.4.2.3 Large Test Problems

The large test problems contain 20 product families, which leads to 80 products in total for the instances with a low number of products per family, and to 200 products in total for the instances with a high number of products per family. The number of machines on each of the three stages is displayed in Table 6.9. On average, a total of 2 000 jobs have to be scheduled.

Table 6.12 shows the results. They are very similar to the ones for the small test problems. There are some minor differences for the relative flow time: The effect of the number of machines and the family demand pattern is only significant at a 5% error level. Further, the mean relative flow time remains higher for a high capacity utilization, but for the large test instances, the effect is no longer statistically significant. A high capacity utilization also seems to

Table 6.12. Factorial design results for large test problems

	#Setups	StdDev	RFT	StdDev	Time
#Products					
low	322.2**	134.29	4.298	3.092	0:01:08**
high	550.0	225.12	4.233	3.116	0:20:51
#Machines					
low	272.4**	91.15	3.880*	2.530	0:06:40**
high	599.7	180.71	4.651	3.546	0:15:19
ProdDemPat					
equal	471.4**	237.30	4.225	2.938	0:11:12
diff sizes	400.8	189.69	4.306	3.262	0:10:47
FamDemPat					
equal	436.1	218.82	4.676*	3.362	0:11:07
diff sizes	436.1	216.59	3.855	2.763	0:10:52
CapUtil					
low	436.5	222.77	3.972	2.772	0:09:58
high	435.6	212.53	4.559	3.379	0:12:01
SetupTime					
low	436.1	223.21	2.085**	0.847	0:12:17*
high	436.0	212.07	6.446	3.000	0:09:42
Total	436.1	217.43	4.266	3.100	0:11:00

slightly increase the computation time, albeit this effect is not significant either. Additionally, high setup times seem to have a positive effect on the computation time, though only significant at a 5% error level.

Comparing the results for the small and the large test problems directly, one can see that the number of setups is higher for the large test problems. This is obvious because more product families and products have to be scheduled (five product families for the small test problems and 20 for the large test problems). With this in mind, the number of setups for the large test problems (on average 436.1 per instance, 21.8 per family) is relatively smaller than the number of setups for the small test problems (on average 130.2 per instance, 26.0 per family).

There is also an increase of the relative flow time for the large test problems. This is in accordance with the findings that the higher the number of machines, the longer the relative flow time: The large test problems contain 2.5 times as many machines as the small test problems.

The average computation time is about three times longer for the large test problems. This is a very good result, as it shows a relatively slow increase of the computation time with the problem size: The large problems contain four times as many product families and 2.5 times as many machines as the small test problems.

6.5 Summary

Phase II has delivered a schedule on product family level. The schedule consists of machine/time slots that determine when and on which machine to produce a product family unit. Phase III assigns an individual product of the respective product family to each of these slots. All production stages are considered simultaneously. The objective is to minimize intra-family setup costs and to keep the low flow time of Phase II.

Two nested Genetic Algorithms are employed to solve the problem. They consider the product families consecutively. An outer Genetic Algorithm calculates a target number of setups for each product family. This figure is used to derive a target batch-size and a target number of setups for the individual products. Applying these target values, an inner Genetic Algorithm determines a schedule on product level for all production stages.

Extensive computational tests have been performed to find 'optimal' setting-parameters and to identify problem characteristics that influence the solution quality and the computation time of the procedures. The tests indicate that the procedures are able to solve relatively large problems in a short or moderate amount of time. Moreover, the computation time seems to increase only slowly with the number of product families and the number of machines. This allows the use of the procedures in real business applications. The procedures can be adjusted with a lookahead-parameter, which enables a human planner to effectively manage the trade-off between the objectives of low setup costs and low flow time.

At the end of Phase III, a feasible and 'optimal' schedule for the initial lot-sizing and scheduling problem for flexible flow lines has been found. It can be handed over to the shop floor.

7

An Illustrative Example

An opportunity awaits you... in the form of an extra-ball!

In this chapter, we present and solve a small, illustrative example in order to facilitate the understanding of the developed algorithms. For this reason, we show and analyze the specific results after each of the three phases of the solution approach.

7.1 Data of the Example

We consider the following example: Two product families have to be produced in two periods on three production stages. There are two machines on the first stage, four machines on the second, and two machines on the third stage. The first product family (C) consists of three products, $C1$, $C2$ and $C3$. The second product family (D) consists of one product $(D1)$. To keep the example simple, both families share the same process and setup times as well as inventory, back-order and setup costs. Process times are one time unit on stages 1 and 3, and two time units on stage 2. Setup times (for product family setups) are two time units on stages 1 and 2, and five time units on stage 3. Inventory costs are one unit, back-order costs two units, and (system-wide) setup costs for each setup on the bottleneck stage 50 units. Initial back-order and inventory volumes are zero. Tables 7.1 and 7.2 summarize these parameters using the notation from the previous chapters. On each of the stages, half of the machines (i.e. one machine on stage 1, two machines on stage 2, and one machine on stage 3) are initially set up for family C (product $C1$), and the other half for family D. The demand is depicted in Table 7.3. It is shown for the individual products as well as summed up for product family C. A period consists of 12 time units.

Table 7.1. Process and setup times of the example, $p = C, D$

Parameter	Stage $l = 1$	Stage $l = 2$	Stage $l = 3$
M_l	2	4	2
t_{lp}^s	2	2	5
t_{lp}^u	1	2	1

Table 7.2. Costs and initial volumes of the example, $p = C, D$

Parameter	Value
c_p^i	1
c_p^b	2
c_p^s	50
b_p^0	0
y_p^0	0

Table 7.3. Demand volumes of the example

Parameter	Period $t = 1$	Period $t = 2$
$d_{C1,t}$	5	4
$d_{C2,t}$	5	4
$d_{C3,t}$	5	4
$d_{C,t}$	15	12
$d_{D1,t}$	6	8

7.2 Solution after Phase I

Phase I solves the lot-sizing and scheduling problem on product family level for the bottleneck stage. The cumulative process times divided by the available machine capacity is equal for all stages. However, setup times are higher on stage 3. For that reason, we consider stage 3 as the bottleneck stage. The resulting schedule is displayed in Fig. 7.1.

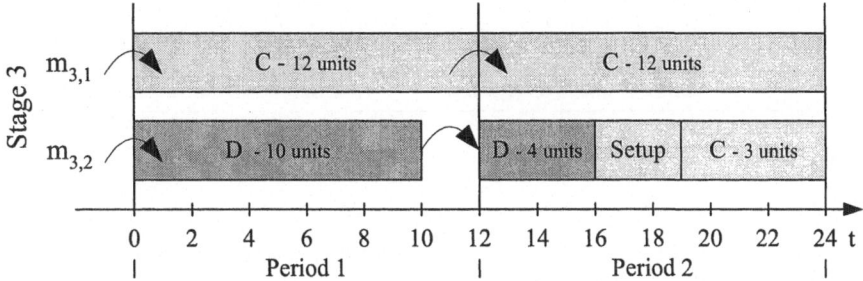

Fig. 7.1. Result after Phase I

As can be seen, machine $m_{3,1}$ (denoting machine 1 on stage 3) produces product family C in period 1 and 2. In period 1, it uses the initial setup state and carries it over to period 2. Machine $m_{3,2}$ also uses its initial setup state in period 1 and produces family D. The setup state is carried over to period 2, where production of family D is continued. Finally, the machine is set up for production of family C.

Comparing the production with the demand volumes, one can see that after period 1, there is an inventory of four units for product family D, while three units of product family C are back-ordered. Any other solution with lower inventory or back-order volumes would incur an additional setup. Besides being unattractive because of the high setup costs of 50 units, an additional setup would consume five capacity units, whereas only two are available. The solution is optimal, as can be proven by the direct CLSPL-BOPM implementation.

7.3 Solution after Phase II

In Phase II, the product family schedule is rolled out to the other production stages. To achieve the same throughput as the bottleneck stage, we have to use one machine on stage 1 and two machines on stage 2 for each machine on stage 3. The result of Phase II is presented in Fig. 7.2. It consists of feasible

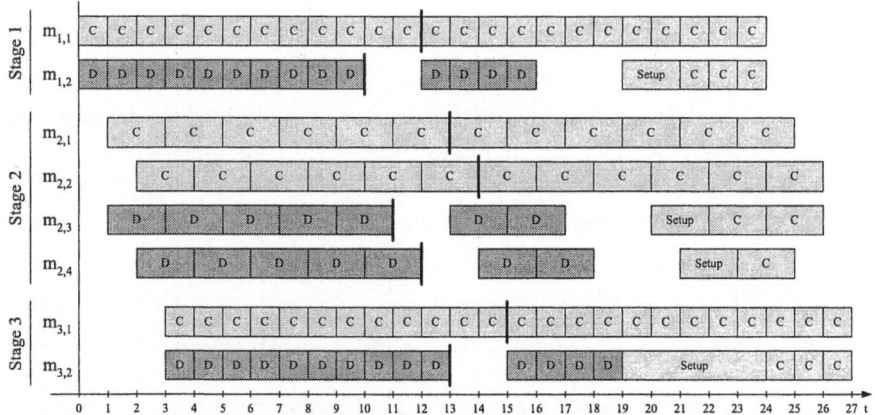

Fig. 7.2. Result after Phase II

machine/time slots that take the precedence constraints between stages into account. The short, vertical bars (for example at time 12 for machine $m_{1,1}$) indicate the period boundaries. More precisely, they show to which period a production slot belongs. As mentioned in Chap. 5, the plan is stretched like an accordion between stages. This makes it impossible to fix a certain period boundary that holds true for all stages.

Besides the necessary postponements that originate from the precedence constraints, one can easily see that the structure of the schedule as shown in Fig. 7.1 remains unchanged. Considering the result in more detail, we notice that all machines producing family D have some idle time in period 1. As it is of no importance when, in a period, a product unit is produced, the idle time is inserted at the end of the period. Further, Phase I has determined that at a certain time in period 2, both machines of stage 3 shall produce family C. This leads to setups on stages 1 and 2 in order to achieve the same throughput. Before being set up, the machines of stages 1 and 2 remain idle for three time units because their setup consumes only two time units, whereas a setup on stage 3 consumes five time units. In this example, the setups are performed as late as possible. However, they may be performed anytime during the idle time.

7.4 Solution after Phase III

Phase III uses the machine/time slots from Phase II and assigns a product unit to each of the product family slots. As in the previous chapter, we assume that the third stage does not differentiate between the products of a family. Thus, on this stage, no setup has to be performed when changing between

Table 7.4. Comparison of performance criteria

Solution	RFT	Intra-family setups
Lookahead 0	1.00	13
Lookahead 6	1.39	10

products of a family and the job sequence has no impact on any performance criteria. It has therefore not been optimized. However, it is displayed in the following figures for the sake of completeness.

The Genetic Algorithms employed in Phase III make use of additional parameters: The length of a sub-period is set to 12 time units, i.e. each period consists of a single sub-period. The other parameters (except the lookahead) are relatively unimportant because of the small size of the example: The Genetic Algorithms are able to explore the complete search space even for small parameter settings. For the outer Genetic Algorithm, the population size is set to 10 with a maximum of 100 generations. For the inner Genetic Algorithm, these parameters are both set to 20. The mutation probability of the inner Genetic Algorithm is 50%. Intra-family setup costs are identical for stages 1 and 2.

Figure 7.3 shows the final solution for a lookahead-parameter of zero. As can be seen, there is no adjustment of the machine/time-slots from Phase II. The average relative flow time (RFT) of this solution is one because no job has to wait between stages. 13 intra-family setups have to be performed (not counting the ones coinciding with product family setups): six on the first stage and seven on the second.

Figure 7.4 presents the solution for a lookahead-parameter of six. In this case, the machine/time-slots from Phase II can no longer be adhered to. Certain operations of a job have to wait for the job to finish on the preceding stage. As a result, the last unit is completed at time 31 (makespan) instead of 27 in Fig. 7.3. Moreover, the relative flow time is increased to 1.39. However, the number of setups has decreased: In total, only 10 intra-family setups have to be performed, four on stage 1 and six on stage 2. This is a reduction of about 23% compared to a lookahead of zero. Table 7.4 summarizes the performance criteria for the two solutions.

Comparing the solutions in more detail, one can easily recognize a different pattern in period 1. On stage 1, both solutions schedule all units of $C1$ before $C2$ and $C3$. This leads to the minimal number of setups for stage 1. However, with a lookahead of zero, this is only possible when both associated machines on stage 2 are loaded with all three products. Thus, both machines have to perform a setup for product $C2$ and $C3$, leading to four setups. With a lookahead of six, each of those machines produces only two different products—

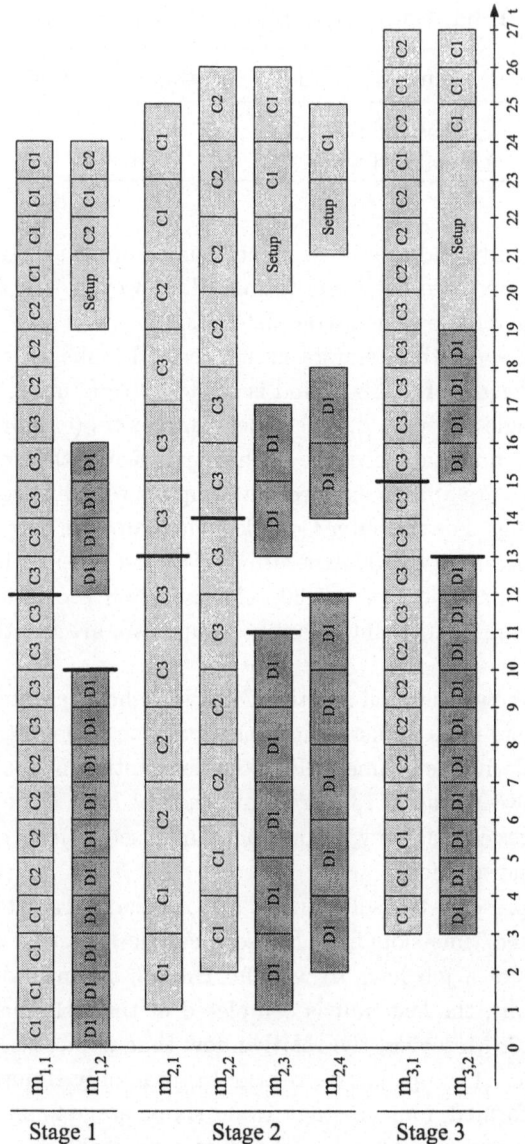

Fig. 7.3. Result after Phase III (*lookahead* = 0)

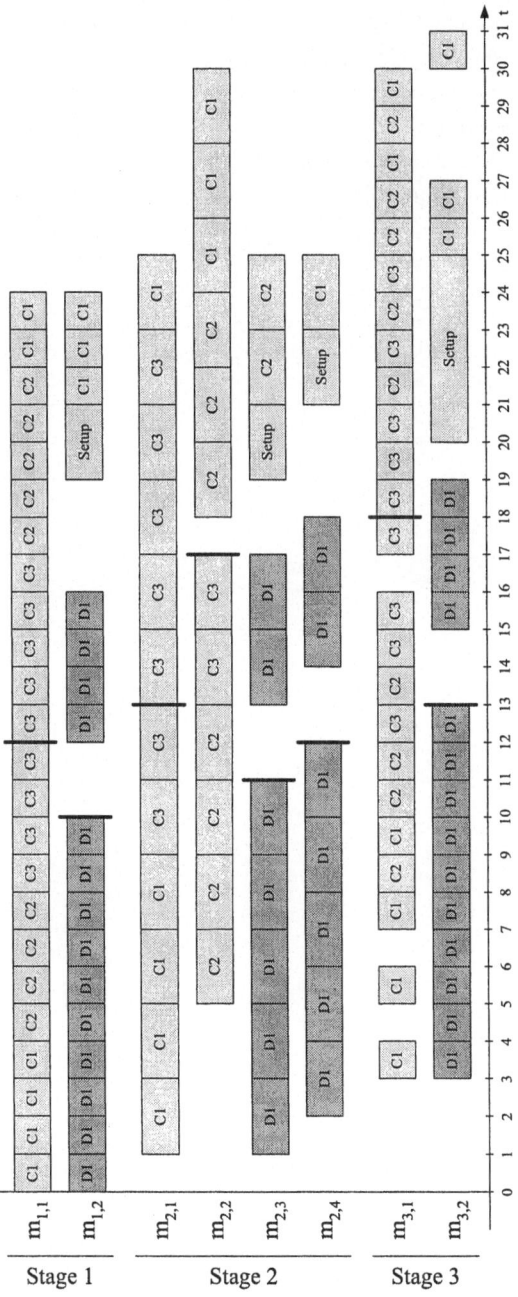

Fig. 7.4. Result after Phase III (*lookahead* = 6)

machine $m_{2,1}$ produces $C1$ and $C3$, machine $m_{2,2}$ produces $C2$ and $C3$, which leads to only three setups (machine $m_{2,1}$ is initially set up for product $C1$). In this way, the lookahead-parameter of six makes it possible to reduce the number of setups on stage 2.

We have also solved the example with a lookahead of 12 time units. However, the solution is identical to the one displayed for a lookahead of six time units.

8

An Application in the Semiconductor Industry

It seems that one of life's payoffs is ahead!

8.1 Semiconductor Process Flow

8.2 Problem Specifics and Dimensions

8.3 Modeling Assumptions

8.4 Short-Term Production Planning

8.5 Medium-Term Capacity Planning

The semiconductor industry manufactures packaged integrated circuits (ICs), commonly known as computer chips. We consider the production of logic chips—as opposed to memory chips. Logic chips perform several logical operations and are embedded in a vast number of applications, such as mobile phones, washing machines or music-players. With the upcoming trend of embedded computing, logic chips will become even more widespread.

8.1 Semiconductor Process Flow

The semiconductor manufacturing process is usually divided into four main production stages: wafer fabrication, probe, assembly, and final test. Wafer fabrication and probe constitute the front-end, where a so-called 'wafer' is produced. A wafer is a thin disk of silicon and contains several ICs. During wafer probe, the wafer is tested and defective ICs are marked with black ink. The back-end, which combines the assembly and test operations, is often physically located at a different place, frequently even on another continent. In the assembly steps, the wafer is sawn and the individual chips are produced. The following test-phase inspects the individual chips before they are shipped to the customer. Figure 8.1 shows an overview of the process flow.

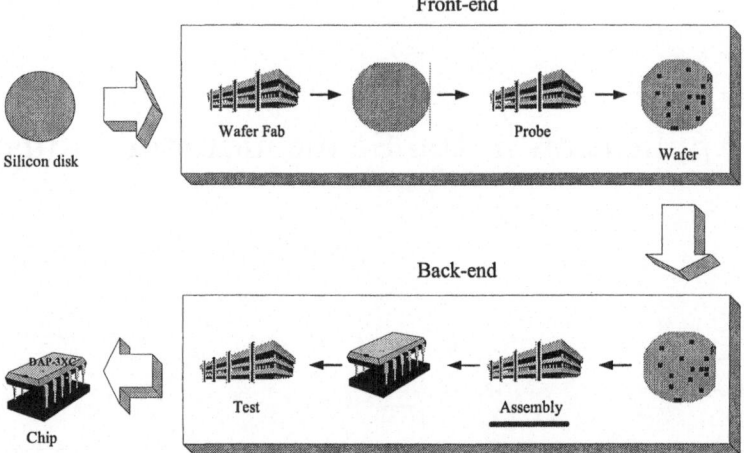

Fig. 8.1. Semiconductor process flow

We consider the back-end assembly, which is typically organized as a flexible flow line. Its main production stages are die attach, wire bonding and molding. After sewing the wafers, the die attach operation mounts the individual ICs on lead frames. The wire bonding process attaches ultra-thin golden wires between each bonding pad on the IC and a connector of the lead frame, to create the electrical path between the IC and the lead fingers. The following molding operation—often also called 'encapsulation'—encloses the individual ICs in plastic or ceramic packages to protect them from the environment. After molding, subsequent production stages include trim&form, where the chips are severed from the lead frame and the lead fingers are formed to become the chip's legs. Finally, the separated chips are handed over to the test operations. Figure 8.2 illustrates the back-end process flow with focus on die attach, wire bonding and molding.

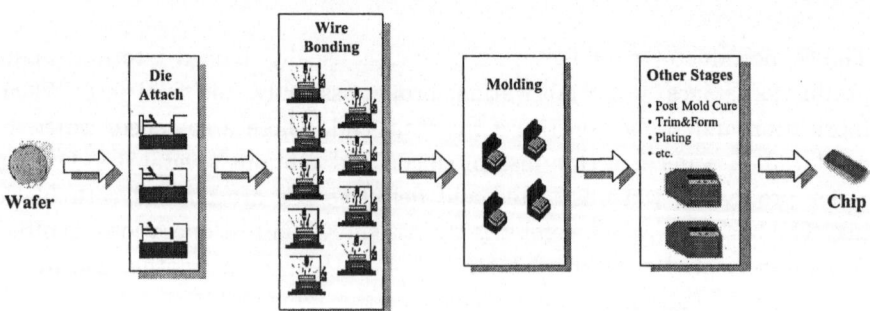

Fig. 8.2. Semiconductor back-end process flow

Only little research is available for back-end facilities. This is probably due to the fact that—from a technological point of view—the manufacturing process in the front-end is more complex. Nevertheless, from a logistical point of view, there are interesting and difficult problems in back-end facilities as well. Papers by Domaschke et al. (1998), Sivakumar and Chong (2001), Yang et al. (2002), Kuhn and Quadt (2002), Quadt and Kuhn (2003) and Quadt and Kuhn (2004), among others, specifically consider back-end problems.

8.2 Problem Specifics and Dimensions

For technical reasons, a certain number of ICs is grouped into batches that cannot be split up during the production process. Such a batch forms the smallest possible production unit, a job.

Logic chips can be differentiated by their visible design. For example, a chip may have contact legs on only two edges or on all four. The number of legs may be different as well. There are also chips that have no legs at all but contact balls on the lower side. The design of a chip is called a 'package'. A package can be identified by simply looking at the final chip. However, the inner logics may be different. They form the so-called 'device'. Each package comprises a number of devices. On the die attach and wire bonding operations, the process times differ slightly between devices of the same package. However, the differences are relatively small compared with the ones between packages. The molding operation does not differentiate between the devices of a package at all—for a given package, the encapsulation process is always the same. Thus, we assume that on all stages, the process time only depends on the package. For the lot-sizing and scheduling problem, the packages form product families and the devices represent individual products.

The job process time differs significantly between production stages. To reach similar production rates, the number of machines per stage also varies significantly. There are about eight times as many wire bonding machines as die attachers or molders because the wire bonding stage needs a relatively long process time.

When changing from one product to another, a setup time is incurred. Setup times are relatively long between product families (major setups, up to 12 hours) and short when changing within a product family (minor setups, ca. one hour). Setup times also vary significantly between stages. On the molding stage, no setup times are incurred when changing products within a family. Setups are in part performed manually. A specifically trained operator has to set up the machine. Thus, a setup does not only consume capacity, but also incurs (labor) costs. Moreover, from a logistical point of view, the setup time is not deterministic.

We focus on three main production stages. Other stages are not incorporated because they do not constitute a planning problem as there is excessive capacity on these stages. In our practical case, the number of machines on the three stages is about 25, 240, and 30, respectively. The number of considered product families is about 20 with each family combining between one and 20 active products. On average, more than 110 products per period are considered in total. The planning period comprises the demand for one month, which is divided into eight periods of half a week. On average, more than 500 jobs have to be scheduled in each period, resulting in an average number of 4 000 jobs in total. Production runs 24 hours a day and seven days a week.

8.3 Modeling Assumptions

For modeling purposes, we simplify the situation in the following aspects: In reality, product family setup times are sequence-dependent, but the degree of sequence-dependency is moderate. It stems from the fact that product families can be grouped to product family types: When changing a machine to another product family type, the setup time is generally longer than within the product family type. As this effect is relatively small compared with the total setup time, the assumption of sequence-independent setup times does not seem to be too restrictive. Thus, we calculate average product family setup times. Product families of different types may also be combined if their process times and setup times to other product families are similar. We further assume that minor setup times can be modeled as a prolongation of the process time. Hence, an intra-family product-to-product setup does not consume time on its own, but this time is added on an average basis to a job's process time. Since, in reality, many stochastic factors such as machine breakdowns and operator availability have an impact on process and setup times (and hence on schedule-fulfillment), this assumption does not seem to inhibit the applicability of the approach.

Another simplification is the assumption of identical machines on the bottleneck stage: In a semiconductor back-end, the machines of a production stage can be clustered into machine-groups. Each machine-group is able to produce a specific (sub-)set of product families at a certain speed. In our approach, we assume that all machines can produce all product families and that the process times depend on the product family only and not on the machine. This restriction does not prevent our approach from being used in a practical setting, for two reasons: Firstly, many real-world machine-groups are in fact physically identical machines that have just been separated by a manual planner to allow a segmentation of the problem. These pseudo machine-groups can be combined into one. Secondly, in the case of physically unidentical machines, the problem may often be clustered by solving it for each of the machine-groups separately. This is possible, because often the subsets of product families that can be produced on different machine-groups do not overlap.

As the intermediate products are sufficiently small, the buffers between stages can be assumed to have infinite capacity. Transport times between stages are negligible.

8.4 Short-Term Production Planning

The presented lot-sizing and scheduling approach is embedded in a decision support system for the described semiconductor back-end facility. The system is used to schedule the production program on a half-weekly basis. After having successfully installed the system, the company has experienced a tremendous decrease of the mean flow time. Flow time is measured as relative flow time (RFT), which is flow time divided by raw process time. In a comparative study with the performance of the real shop floor that has been planned manually, the system led to an RFT reduction of 55%, while at the same time, only 80% of the capacity on the bottleneck stage has been utilized.

8.5 Medium-Term Capacity Planning

The decision support system has also been used in a capacity planning study. The company wants to change the production process in order to use a new product design. For the duration of the re-organization, the customers have the choice between an old and a new variant. Hence, the number of product variants will be doubled as soon as the re-organization process begins. When the re-organization is finished, the number of variants will drop back to its original level as only the new variants are produced. The question is, what impact, if any, the re-organization—i.e. the doubling of the number of product variants and the successive adjustment of production volumes—will have on the relevant performance criteria and if additional production capacities will be required to fulfill the given demand volumes?

The study has been conducted with real business data. The distribution of production volumes between the old and the new variants was varied as shown in Table 8.1.

A distribution of 100–0 means that only old variants are produced. A distribution of 90–10 means that 90% (10%) of the total volume of a product will be 'old' ('new') volume, accordingly for the other distributions. Since the

Table 8.1. Distribution of production volumes

distribution	100–0	90–10	80–20	70–30	60–40	50–50
old	100%	90%	80%	70%	60%	50%
new	0%	10%	20%	30%	40%	50%

distribution of volume is valid for each individual product, in all the distributions from 90–10 to 50–50, there are two times as many product variants as in the 100–0 distribution. The process plans including process times are the same for the respective old and new variants. Hence, for the 40–60, 30–70, 20–80, 10–90 and 0–100 distributions, the results will be analogous and need not to be analyzed.

Figure 8.3 shows the effect of the re-organization on the relative flow time (RFT). The company had predicted that the RFT would increase with more balanced 'old' and 'new' volumes, because of a higher number of product variants. However, the study showed the opposite result. The RFT remained relatively stable and even decreased slightly for more balanced production volumes. The reason is that the number of variants is the same for all mixed distributions, only the respective volumes are different. Furthermore, one could argue that the more balanced the volumes, the fewer the number of low-volume products. This makes the scheduling problem easier and allows a slight decrease of the RFT. Nevertheless, this argumentation cannot explain the relatively high RFT for the 100–0 distribution that only includes 'old' volumes.

Figure 8.4 shows the number of setups on one production stage. The study included 213 parallel machines on this stage and a planning horizon of three days. As soon as the number of variants is increased, the number of setups increases as well. However, the mix of production volumes does not seem to have an effect. For all distributions combining old and new variants, the number of setups remains at a similar level.

As a summary, the planning system has helped the company to determine how many additional machines are needed in order to produce the expected demands and how the relative performance criteria will evolve during the re-organization.

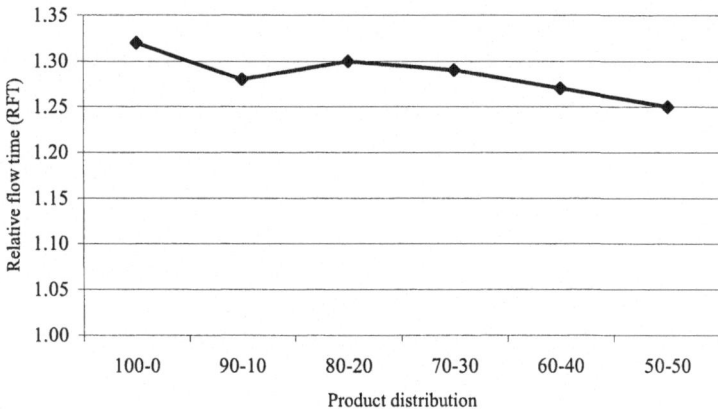

Fig. 8.3. Effect of the production process re-organization on the relative flow time

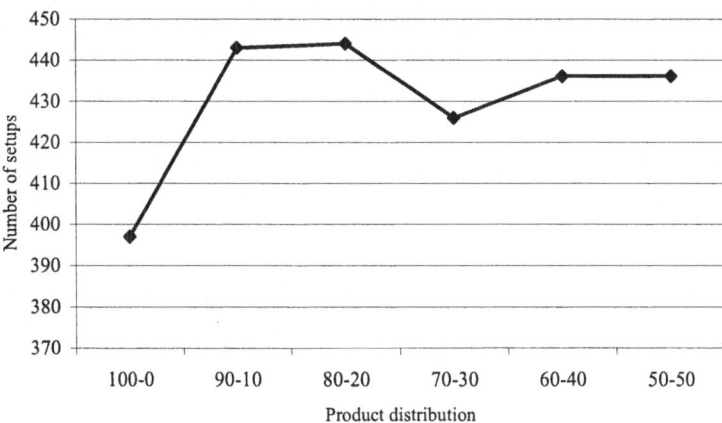

Fig. 8.4. Effect of the production process re-organization on the number of setups

9

Summary and Conclusions

You have come to the end of your journey!

The trend towards a globalized economy leads to an increasing pressure of competition. Many companies face this competition by trying to better fulfill the needs of their customers. To accomplish this goal, they offer a higher number of product variants and shorten delivery times. Both changes require elaborate production planning systems. One can distinguish long-term, medium-term and short-term production planning. Lot-sizing and scheduling are an integral part of short-term production planning.

This study has considered the lot-sizing and scheduling problem for flexible flow line production facilities. Flexible flow lines are flow lines with parallel machines on some or all production stages. They can be found in a vast number of industries. The goal of a lot-sizing and scheduling routine is to determine production quantities, machine assignments and product sequences. The latter two represent a schedule, while the former are the result of a lot-sizing procedure. We have identified several important features in industrial practice that are relevant for lot-sizing and scheduling of flexible flow lines. Mainly, these are the inclusion of back-orders, the aggregation of products to product families and the explicit consideration of parallel machines. Lot-sizing and scheduling decisions are interdependent: It is not possible to determine a schedule without knowing the lot-sizes that have to be produced. On the other hand, one cannot calculate optimal lot-sizes without knowing the machine assignments and the product sequences. These interdependencies make it impossible to effectively solve the problems separately. However, an in-depth literature review revealed that there seems to be no solution procedure for flexible flow lines that considers the lot-sizing and scheduling problem integratively. We have presented such an integrative solution approach for the combined lot-sizing and scheduling problem. The objective is to minimize setup, inventory holding and back-order costs as well as mean flow time through the system.

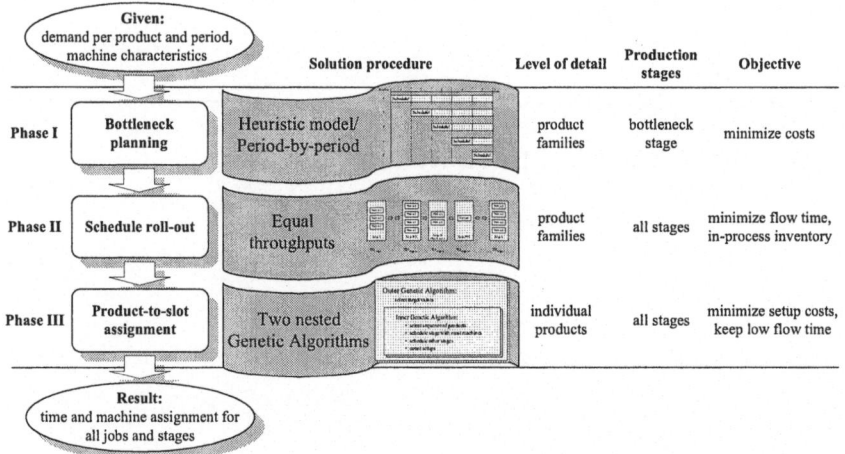

Fig. 9.1. Summary of the solution approach and the developed procedures

In the remainder, we give, firstly, a summary of the solution approach. Secondly, we elaborate on its integration into hierarchical production planning systems and rolling planning horizon environments as well as discuss its usability in real business applications. Finally, we indicate areas that are open for further research.

The **solution approach** consists of the phases 'Bottleneck planning', 'Schedule roll-out' and 'Product-to-slot assignment'. Figure 9.1 shows an overview and depicts the developed procedures. The approach is bottleneck-oriented. A human planner is usually aware which production stage constitutes the bottleneck. If this is not the case, the bottleneck stage has to be determined beforehand.

The first phase of the solution approach solves a single stage lot-sizing and scheduling problem on the bottleneck stage. Products are aggregated to product families. The suggested solution procedure is based on a new 'heuristic' model formulation that uses integer variables in contrast to binary variables as employed by standard formulations. This makes it possible to solve the model to optimality or near-optimality using standard algorithms that are embedded in commercial software packages (e.g. CPLEX). However, the heuristic model cannot capture all capacity restrictions as imposed by the original problem. For this reason, it is embedded in a period-by-period heuristic. In each of its iterations, an instance of the model is solved using standard algorithms (CPLEX). In the first iteration for period 1, all periods are included in the model. In later iterations for period t, only the periods from t to the planning horizon are covered. A scheduling sub-routine is invoked everytime an instance of the model has been solved. It tries to schedule the production volumes of the actual period as given by the solution of the model. In this way, a possible capacity misestimation is detected and—in the case of an overestimation—the surplus production volumes are postponed to subsequent periods. When there is a remaining production volume at the end of the planning horizon, a period backtracking is performed and the respective volumes are shifted towards the beginning of the planning interval.

The second phase generates a product family plan on all production stages. Thus, it keeps the aggregation level of product families, but explicitly includes the non-bottlenecks stages. It assigns a production time and a machine to each product family unit, which means that it solves a multi-stage scheduling and batching problem. However, the number of machines per product family and the batch-sizes are pre-determined by the first phase. The objective is to minimize mean flow time. This implies a minimization of waiting times between stages, and hence of in-process inventory. As a side effect, the usage of intermediate buffers is also minimized. The basic idea of the solution procedure is to establish the same product family throughput on all stages. In this way, each preceding stage supplies the intermediate products just in time for the subsequent stage. Thus, the products do not have to wait between stages and the flow time as well as the inventory volumes of intermediate products are minimized. The result of Phase II are so-called 'machine/time slots', signalling at what time and on which machine a product family unit is to be produced.

The third phase considers the individual products and calculates the final schedule, which consists of exact production times and machine assignments for every product unit on all stages. The problem is to determine how many and which machines to set up and when to produce the individual product units. Hence, it is a scheduling and batching problem for all stages on product level. The solution is generated using the machine/time slots from the second phase: We assign an individual product unit of the respective family to each of the slots under the objective of minimizing intra-family setup costs, while keeping the low flow time of the second phase. There is a trade-off between the two objectives. Two nested Genetic Algorithms are employed to solve the problem. They consider the product families separately. The stage with the largest number of parallel machines appears to be the most difficult stage. Therefore, it is considered first. An outer Genetic Algorithm sets a target number of setups and a target batch-size for each product of the actual family. With these target values, the problem resembles a two-dimensional packing problem: On a Gantt-Chart, the available slots for the product family form a pallet and the products form rectangular boxes that have to be placed on the pallet. An inner Genetic Algorithm tries to find an optimal packing pattern. The packing pattern may in turn be interpreted as a schedule. To schedule the other stages, we generate a job sequence over all parallel machines. The other stages are loaded in this sequence, with some local re-sequencing to achieve low setup costs on these stages as well.

The solution procedures can be embedded into **hierarchical production planning systems**. Over the last few years, several commercial vendors have developed hierarchical production planning systems and their revenues have risen sharply. The study at hand may help improve these systems by delivering specialized routines to solve short-term planning problems for flexible flow lines. In the context of real business applications, **rolling planning horizons** are pre-dominant. Such environments are characterized by the fact that a certain frozen interval of an afore-determined schedule shall not be modified in later planning runs. The presented solution approach seems very well suited for a rolling planning horizon, since the top-level phase is based on a period-by-period procedure. As such, the concept of a frozen time interval is inherent to the approach. Parts of the solution that shall not be overwritten in later iterations may easily be fixed at their current specification.

Extensive computational tests have shown that the developed heuristics outperform standard procedures in both solution quality and computation time. Various factors influencing the performance have been identified and analyzed. The heuristics are able to solve practically sized problems in a moderate amount of time. Thus, they can be used in real **business applications**. They have been embedded in a scheduling system for a semiconductor assembly facility. The system has been developed to run on standard personal computers using Microsoft Windows as the operating system and is mainly used by medium and low-level managers that generate a plan on a half-weekly basis. It has also been used for a capacity planning study. A user-friendly interface allows the planner to easily change certain data and perform what-if analyses. The resulting schedule can either be plotted as a Gantt-Chart or displayed as a detailed schedule in a Microsoft Excel format.

As is usual in such circumstances, such a system is not able to take some minor constraints into account. An example would be the availability of specific tools that are needed for production. However, such constraints may often be neglected as they can be solved outside the system. For example, additional tools may be purchased if their price is negligible. Another approach is to allow the human planner to manually adjust a computed schedule. In our case, this was made possible through the Microsoft Excel output interface.

When implementing a production planning system, it is generally helpful to explain some of the algorithmic concepts to the planners. In this way, they are able to understand why a certain data set leads to a specific solution. This results in a higher degree of trust in the computed schedule and the planning system itself. It is of further importance that the operators on the shop floor are able to read and understand the schedule and its implied flow of material.

Specific training will usually be required. Following a deterministic schedule in a stochastic environment (e.g. machine failures) involves a relatively high intellectual involvement. For example, if two machines have to be set up by a single operator, he has to choose which one to set up first. For this, he must understand which machine is more crucial than the other, such as when one is (or is going to be) behind schedule. Ultimately, one should also consider revising the operators' bonus payment system to support such a way of involvement.

There are several aspects open for **further research**: The developed heuristic seems to be the only solution procedure for the lot-sizing and scheduling problem of flexible flow lines. We have compared the heuristic with standard algorithms as they are embedded in commercial software packages (CPLEX). These standard algorithms are only able to solve relatively small problems. Alternative solution procedures would be useful in order to compare solution quality and computation times. Exact solution procedures, in particular, would be beneficial, as they would allow a comparison of the solutions of the heuristic with provably optimal solutions.

Considering the planning approach itself, one can also identify several areas open for future research: The approach is bottleneck-oriented. This might lead to sub-optimal or even infeasible solutions if more than one production stage constitutes a bottleneck or if the bottleneck moves during the planning interval. In such cases, one should consider using a different approach. However, the approach is quite general and does not depend on the specific procedures employed in each phase. Thus, the developed algorithms may be replaced by others more suitable for different applications. For example, the procedure used in the second phase does not shift production volumes between periods. It takes the volumes of the bottleneck stage and schedules them on the non-bottleneck stages, separately for each period. Hence, it assumes that the complete production volume as given by the bottleneck stage can be produced on all other production stages. While this does not impair its applicability in the illustrated business case, it might lead to problems in others. An alternative procedure could move a production volume of a non-bottleneck stage from one period to another in order to save setup times. However, it should be noticed that the proposed algorithm leads to very low flow times, as it minimizes the waiting time between stages. Thus, as long as it is applicable, it should not be modified. If in doubt, one could also try to develop modifications similar to the presented cyclic production technique, which allows one to keep to the general concept. Incurring only moderately longer flow times and higher setup costs, this technique entails a better capacity utilization and may lead to feasible schedules even though the original procedure would not be applicable in the specific environment.

List of Abbreviations

You want more?

#Best	Number of best solutions found
CapUtil	Capacity utilization
CLSD-BOPM	Capacitated Lot-sizing problem with Sequence-Dependent setups, Back-Orders and Parallel Machines
CLSP	Capacitated Lot-Sizing Problem
CLSPL	Capacitated Lot-Sizing Problem with Linked lot-sizes
CLSPL-BOPM	Capacitated Lot-Sizing Problem with Linked lot-sizes, Back-Orders and Parallel Machines
DemPat	(Product family) demand pattern
DemProb	Demand probability
DemVar	Demand variation
diff sizes	Different sizes demand pattern
FamDemPat	Product family demand pattern
HM	Heuristic Model
HMO	Heuristic Model with Overtime
IC	Integrated circuit (computer chip)
#Inst	Number of instances
InvCosts	Inventory costs
LSP-BOTH	Lot-sizing and Scheduling Procedure combining LSP-NSCF and LSP-WSCF
LSP-BOTH-CA	Lot-sizing and Scheduling Procedure combining LSP-NSCF-CA and LSP-WSCF-CA

LSP-NSCF	Lot-sizing and Scheduling Procedure No Setup Costs Factor
LSP-NSCF-CA	Lot-sizing and Scheduling Procedure No Setup Costs Factor with Capacity Adjustment
LSP-WSCF	Lot-sizing and Scheduling Procedure With Setup Costs Factor
LSP-WSCF-CA	Lot-sizing and Scheduling Procedure With Setup Costs Factor and Capacity Adjustment
Max	Maximal (worst) ratio
Min	Minimal (best) ratio
Mutation prob.	Mutation probability
#Opt	Number of provably optimal solutions found
pos trend	Positive trend demand pattern
ProdDemPat	Product demand pattern
RFT	Relative Flow Time
StdDev	Standard deviation

List of Symbols

See what greed will get you?

APT^{MT}	Average processing time over all machines and periods on the bottleneck stage
APT_l^{MT}	Average processing time over all machines and periods on stage l
APT^{PT}	Average processing time over all periods caused by an average product family on the bottleneck stage
APT_l^{PT}	Average processing time over all periods caused by an average product family on stage l
b	Batch-size for current product
b_{pm}	For Phase II: Batch-size for product family p per cycle on machine m
b_{pt}	Back-order volume of product family p at the end of period t
b_p^0	Initial back-order volume of product family p at beginning of planning interval
B	Bottleneck stage
c_p^b	Back-order costs of product family p
c_p^i	Inventory holding costs of product family p
c^o	Overtime capacity costs
c_p^s	(System-wide) setup costs of product family p
c_{pl}^{si}	Intra-family setup costs of product family p on stage l
C	Capacity of a (parallel) machine per period
C_m	Remaining capacity of machine m after step 2 of the scheduling procedure in Phase I

C^A	Total per period capacity aggregated over all parallel machines on the bottleneck stage, $C^A = M \cdot C$
C_t^A	Total capacity in period t aggregated over all parallel machines on the bottleneck stage
C_p^c	Capacity of a single machine on the bottleneck stage in units of product family p when the setup can be carried over from the previous period (machine-quantity with setup carry-over)
C^f	Additional capacity needed to find a feasible solution without overtime
C_p^f	Additional capacity needed for processing to find a feasible solution without overtime
C_s^f	Additional capacity needed for setups to find a feasible solution without overtime
C_t^{orig}	Original aggregate capacity in period t
C^O	Overtime capacity usage
C_t^{reduced}	Reduced aggregate capacity in period t
C_p^s	Capacity of a single machine on the bottleneck stage in units of product family p when a setup has to be performed (machine-quantity with setup)
C_t^{used}	Capacity usage in period t
CL	Copy length
CP	Copy position
CT_m	Cycle time on machine m
CT^{max}	Maximum cycle time
CT^{min}	Minimum cycle time
CT^s	Shortest possible cycle time
CUF	Capacity utilization factor
d_{pt}	Demand volume of product family p in period t
D_p	Demand volume for product family p (determined by production volume in Phase I)
D_{pk}	Demand volume for product k of product family p (determined by production volume in Phase I)
D_{pkt}	Demand volume for product k of family p in period t

D_{pt}	For computational tests: Demand volume of product family p in period t
D_p^r	Remaining demand volume for product family p
D_{pk}^r	Remaining demand volume for product k of product family p
f	Setup costs devaluation factor
F	Big number, $F \geq \max\left\{\max_{p \in \bar{P}}\left\{t_p^s\right\}, M\right\}$
K_p	Number of products in product family p
l_i	Number of slots after position i having a start-time within the lookahead
lookahead	Lookahead
L	Number of production stages
M	Number of parallel machines on the bottleneck stage, $\bar{M} = \{1 \ldots M\}$
M_l	Number of parallel machines on stage l
M'	Number of remaining (i.e. not yet fully utilized) parallel machines on the bottleneck stage, $\bar{M}' = \{1 \ldots M'\}$
M^{Sp}	Total number of machines for product family p on stage S in the schedule retrieved from Phase II
n_{lp}	Number of machines currently producing product family p on stage l
P	Number of product families, $\bar{P} = \{1 \ldots P\}$
P_m^c	Product families that share machine m in a production cycle
PM	Number of parallel machines to load with current product
PM^u	Number of parallel machines having an unassigned slot for current product family with a starting time in the actual sub-period
PS	Population size (of the outer Genetic Algorithm)
q_p	Remaining production volume of product family p after step 2 of the scheduling procedure in Phase I
r_p	Production rate of product family p
r_{pm}	Production rate of product family p on machine m
s_i	For outer Genetic Algorithm: Individual i of the initial population

s_p	Setup carry-over bonus for product family p ($s_p = c_p^s$, if $t_1 < T$ and the optimal solution of model HMO scheduled a setup for p in $t_1 + 1$)
S	Production stage with the largest number of parallel machines
S_l	Number of setups on stage l
S_p^{LB}	Lower bound on the number of intra-family setups for product family p
S_p^r	Remaining number of intra-family setups for product family p
S_{pk}^r	Remaining number of intra-family setups for product k of product family p
S_p^{UB}	Upper bound on the number of intra-family setups for product family p
t_1	Index of first period in models HM and HMO, $t_0 = t_1 - 1$
t^i	Idle time between two consecutive product family units
t_i^l	Start-time of slot i of the current product family on stage l
t_{lp}^s	Setup time of product family p on stage l
t_p^s	Setup time of product family p on the bottleneck stage
t_{lp}^u	Per unit process time of product family p on stage l
t_p^u	Per unit process time of product family p on the bottleneck stage
$\tilde{t}_{lp}^u$	Preliminary per unit process time of product family p on stage l
T	Number of periods, $\bar{T} = \{1 \ldots T\}$ For models HM and HMO: Index of last period, $\bar{T} = \{t_1 \ldots T\}$
T_p^b	Target batch-size for the products of product family p
T^f	Last period of the frozen interval
T_p^s	Target number of intra-family setups for product family p
T_{pk}^s	Target number of intra-family setups for product k of product family p
u_{pm}	Utilization of the last machine m scheduled for product family p
$U(a; b)$	Uniform distribution of real values between a and b
x_{pm}	For model Scheduler: Fraction of the remaining production volume q_p of product family p produced on machine m
x_{pt}	Production volume of product family p in period t

x_{ptm}	Production volume of product family p in period t on machine m
x_{pt}^D	Dummy production volume of product family p in period t
y_{pt}	Inventory volume of product family p at the end of period t
y_p^0	Initial inventory volume of product family p at beginning of planning interval
z	Big number, $z \geq \sum_{\substack{p \in \bar{P} \\ t \in \bar{T}}} d_{pt}$
γ_{pqm}	Binary setup variable indicating a setup from product family p to product family q on machine m ($\gamma_{pqm} = 1$, if a setup is performed from product family p to product family q on machine m)
γ_{pt}	Binary setup variable for product family p in period t ($\gamma_{pt} = 1$, if a setup is performed for product family p in period t) For models HM and HMO: Number of setups for product family p in period t
γ_{ptm}	Binary setup variable for product family p in period t on machine m ($\gamma_{ptm} = 1$, if a setup is performed for product family p in period t on machine m)
γ_{qptm}	Binary setup variable indicating a setup from product family q to product family p in period t on machine m ($\gamma_{qptm} = 1$, if a setup is performed from product family q to product family p in period t on machine m)
γ_p^i	Binary variable ($\gamma_p^i = 1$, if a setup for product family p is impossible due to its setup time)
$\hat{\gamma}_{pt}$	Number of machines used for product family p in period t with setup carry-over from period $t - 1$
π_i	Product that will be considered at position i by the inner Genetic Algorithm
π_{pm}	Sequencing variable indicating the ordinal position of product family p on machine m. The larger π_{pm}, the later product family p is scheduled on m
π_{ptm}	Sequencing variable indicating the ordinal position of product family p on machine m in period t. The larger π_{ptm}, the later product family p is scheduled on m in t
σ	Interval $\left[S_p^{LB}; S_p^{UB}\right]$

τ First period that the scheduling procedure is not able to schedule the complete production volume given by the solution of model HMO

ζ_{pm} For model Scheduler: Binary setup carry-over variable for product family p on machine m ($\zeta_{pm} = 1$, if the setup state for product family p on machine m is carried over from period t_1 to $t_1 + 1$)

ζ_{pt} Binary linking variable for product family p in period t ($\zeta_{pt} = 1$, if the setup state for product family p is carried over from period t to $t + 1$)

For models HM and HMO: Number of machines with setup carry-over for product family p between periods t and $t + 1$

ζ_{ptm} Binary linking variable for product family p in period t on machine m ($\zeta_{ptm} = 1$, if the setup state for product family p on machine m is carried over from period t to $t + 1$)

ζ_p^0 Initial setup state of product family p at beginning of planning interval ($\zeta_p^0 = 1$, if the machine is initially set up for product family p)

For models HM and HMO: Initial number of machines set up for product family p

ζ_{pm}^0 Initial setup state of product family p on machine m at beginning of planning interval ($\zeta_{pm}^0 = 1$, if machine m is initially set up for product family p)

For model Scheduler: Initial setup state of product family p on machine m after step 2 of the scheduling procedure in Phase I ($\zeta_{pm}^0 = 1$, if machine m is initially set up for product family p)

ζ_{pt}^c Number of machines with setup carry-over for product family p between periods t and $t + 1$ that have been carried over from period $t - 1$

ζ_{pt}^s Number of machines with setup carry-over for product family p between periods t and $t + 1$ that have been set up in period t

References

Not an ordinary game, nor an ordinary player!

Adler, L., N. Fraiman, E. Kobacker, M. Pinedo, J. C. Plotnicoff, and T. P. Wu (1993). BPSS: A scheduling support system for the packaging industry. *Operations Research 41*(4), 641–648.

Agnetis, A., A. Pacifici, F. Rossi, M. Lucertini, S. Nicoletti, F. Nicolo, G. Oriolo, D. Pacciarelli, and E. Pesaro (1997). Scheduling of flexible flow lines in an automobile assembly plant. *European Journal of Operational Research 97*, 348–362.

Allahverdi, A., J. N. Gupta, and T. Aldowaisan (1999). A review of scheduling research involving setup considerations. *Omega – The International Journal of Management Science 27*, 219–239.

Aras, O. A. and L. A. Swanson (1982). A lot sizing and sequencing algorithm for dynamic demands upon a single facility. *Journal of Operations Management 2*(3), 177–185.

Aytug, H., M. Khouja, and F. E. Vergara (2003). Use of genetic algorithms to solve production and operations management problems: A review. *International Journal of Production Research 41*(17), 3955–4009.

Azizoglu, M., E. Cakmak, and S. Kondakci (2001). A flexible flowshop problem with total flow time minimization. *European Journal of Operational Research 132*(3), 528–538.

Bahl, H. C., L. P. Ritzman, and J. N. D. Gupta (1987). Determining lot sizes and resource requirements: A review. *Operations Research 35*(3), 329–345.

Baker, T. and J. Muckstadt, Jr. (1989). The CHES problems. Working Paper, Chesapeake Decision Sciences, Inc., 200 South Street, New Providence, NJ 07974, USA.

Barr, R. S., B. L. Golden, J. P. Kelly, M. G. C. Resende, and W. R. Stewart (1995). Designing and reporting on computational experiments with heuristic methods. *Journal of Heuristics 1*(1), 9–32.

Batini, C., S. Ceri, and S. B. Navathe (1992). *Conceptual Database Design: An Entity-Relationship Approach.* Redwood City, CA: Benjamin/Cummings.

Billington, P. J., J. O. McClain, and L. J. Thomas (1983). Mathematical programming approaches to capacity-constrained MRP systems: Review, formulation and problem reduction. *Management Science 29*(10), 1126–1141.

Bitran, G. R. and H. H. Yanasse (1982). Computational complexity of the capacitated lot size problem. *Management Science 28*(10), 1174–1186.

Blackburn, J. D. and R. A. Millen (1980). Heuristic lot sizing performance in a rolling-schedule environment. *Decision Sciences 11*, 691–701.

Blazewicz, J., K. H. Ecker, E. Pesch, G. Schmidt, and J. Weglarz (2001). *Scheduling Computer and Manufacturing Processes* (2nd ed.). Berlin: Springer.

Botta-Genoulaz, V. (2000). Hybrid flow shop scheduling with precedence constraints and time lags to minimize maximum lateness. *International Journal of Production Economics 64*, 101–111.

Brah, S. A. and J. L. Hunsucker (1991). Branch and bound algorithm for the flow shop with multiple processors. *European Journal of Operational Research 51*(1), 88–99.

Brah, S. A. and L. L. Loo (1999). Heuristics for scheduling in a flow shop with multiple processors. *European Journal of Operational Research 113*(1), 113–122.

Brandimarte, P. (1993). Routing and scheduling in a flexible job shop by tabu search. *Annals of Operations Research 41*, 157–183.

Brockmann, K. and W. Dangelmaier (1998). A parallel branch & bound algorithm for makespan optimal sequencing in flow shops with parallel machines. In P. Borne, M. Ksouri, and A. El Kamel (eds.), *Proceedings of the IMACS Multiconference on Computational Engineering in Systems Applications (CESA 1998)*, Volume 3, pp. 431–436. Hammamet, Tunisia.

Brockmann, K., W. Dangelmaier, and N. Holthöfer (1997). Parallel branch & bound algorithm for makespan optimal scheduling in flow shops with multiple processors. In P. Kischka, H.-W. Lorenz, U. Derigs, W. Domschke, P. Kleinschmidt, and R. Möhring (eds.), *Operations Research Proceedings 1997*, pp. 428–433. Berlin: Springer.

Campbell, H. G., R. A. Dudek, and M. L. Smith (1970). A heuristic algorithm for the n job, m machine sequencing problem. *Management Science 16*(10), B-630–B-637.

Carlier, J. and E. Néron (2000). An exact method for solving the multiprocessor flow-shop. *RAIRO Operations Research 34*, 1–25.

Cheng, C. H., M. S. Madan, Y. Gupta, and S. So (2001). Solving the capacitated lot-sizing problem with backorder consideration. *Journal of the Operational Research Society 52*, 952–959.

Cheng, T. C. E., J. N. D. Gupta, and G. Wang (2000). A review of flowshop scheduling with setup times. *Production and Operations Management 9*(3), 262–282.

Dauzère-Pérès, S. and J. Paulli (1997). An integrated approach for modeling and solving the general multiprocessor job-shop scheduling problem using tabu search. *Annals of Operations Research 70*, 281–306.

de Matta, R. and M. Guignard (1995). The performance of rolling production schedules in a process industry. *IIE Transactions 27*(5), 564–573.

Derstroff, M. C. (1995). *Mehrstufige Losgrößenplanung mit Kapazitätsbeschränkungen*. Heidelberg: Physica (in German).

Diaby, M., H. C. Bahl, M. H. Karwan, and S. Zionts (1992). A lagrangean relaxation approach for very-large-scale capacitated lot-sizing. *Management Science 38*(9), 1329–1340.

Dillenberger, C., L. F. Escudero, A. Wollensak, and W. Zhang (1993). On solving a large-scale resource allocation problem in production planning. In G. Fandel, T. Gulledge, and A. Jones (eds.), *Operations Research in Production Planning and Control*, pp. 105–119. Berlin: Springer.

Dillenberger, C., L. F. Escudero, A. Wollensak, and W. Zhang (1994). On practical resource allocation for production planning and scheduling with period overlapping setups. *European Journal of Operational Research 75*(2), 275–286.

Dimopoulos, C. and A. M. S. Zalzala (2000). Recent developments in evolutionary computation for manufacturing optimization: problems, solutions, and comparisons. *IEEE Transactions on Evolutionary Computation 4*(2), 93–113.

Ding, F. Y. and D. Kittichartphayak (1994). Heuristics for scheduling flexible flow lines. *Computers & Industrial Engineering 26*, 27–34.

Dixon, P. and E. Silver (1981). A heuristic solution procedure for the multi-item single-level limited capacity lot-sizing problem. *Journal of Operations Management 2*(1), 23–39.

Dogramaci, A., J. E. Panayiolopoulos, and N. R. Adam (1981). The dynamic lot sizing problem for multiple items under limited capacity. *AIIE Transactions 13*(4), 294–303.

Domaschke, J., S. Brown, J. Robinson, and F. Leibl (1998). Effective implementation of cycle-time reduction strategies for semiconductor backend manufacturing. In D. J. Medeiros, E. F. Watson, J. S. Carson, and M. S. Manivannan (eds.), *Proceedings of the 1998 Winter Simulation Conference*, Volume 2, pp. 985–992. IEEE, Piscataway, NJ.

Drexl, A., B. Fleischmann, H.-O. Günther, H. Stadtler, and H. Tempelmeier (1994). Konzeptionelle Grundlagen kapazitätsorientierter PPS-Systeme. *Zeitschrift für betriebswirtschaftliche Forschung 46*(12), 1022–1045 (in German).

Drexl, A. and A. Kimms (1997). Lot sizing and scheduling – Survey and extensions. *European Journal of Operational Research 99*, 221–235.

Eppen, G. D. and R. K. Martin (1987). Solving multi-item capacitated lot-sizing problems using variable redefinition. *Operations Research 35*(6), 832–848.

Fleischmann, B. and H. Meyr (1997). The general lotsizing and scheduling problem. *OR Spektrum 19*, 11–21.

Fleischmann, B. and H. Meyr (2003). Planning hierarchy, modeling and advanced planning systems. In A. G. de Kok and S. C. Graves (eds.), *Handbooks in Operations Research and Management Science – Supply Chain Management: Design, Coordination and Operation*, Volume 11, pp. 457–523. Amsterdam: Elsevier.

Fogel, L. J., A. J. Owens, and M. J. Walsh (1966). *Artificial Intelligence Through Simulated Evolution*. New York, NY: Wiley.

Gao, Y. (2000). A heuristic procedure for the capacitated lot sizing problem with set-up carry-over. *Control Theory and Applications 17*(6), 937–940.

Garey, M. R. and D. S. Johnson (1979). *Computers and Intractibility: A Guide to the Theory of NP-Completeness*. San Francisco, CA: Freeman and Co.

Garey, M. R., D. S. Johnson, and R. Sethi (1976). The complexity of flowshop and jobshop scheduling. *Mathematics of Operations Research 1*(2), 117–129.

Gelders, L. F., J. Maes, and L. N. Van Wassenhove (1986). A branch and bound algorithm for the multi-item single level capacitated dynamic lotsizing problem. In S. Axsäter, C. Schneeweiss, and E. Silver (eds.), *Multistage Production Planning and Inventory Control*, pp. 92–108. Berlin: Springer.

Goldberg, D. E. (2003). *Genetic Algorithms in Search, Optimization, and Machine Learning*. 1989. Reprint, Boston, MA: Addison-Wesley.

Gopalakrishnan, M. (2000). A modified framework for modelling set-up carryover in the capacitated lotsizing problem. *International Journal of Production Research 38*(14), 3421–3424.

Gopalakrishnan, M., K. Ding, J.-M. Bourjolly, and S. Mohan (2001). A tabu-search heuristic for the capacitated lot-sizing problem with set-up carry-over. *Management Science 47*(6), 851–863.

Gopalakrishnan, M., D. Miller, and C. Schmidt (1995). A framework for modelling setup carryover in the capacitated lot sizing problem. *International Journal of Production Research 33*, 1973–1988.

Graham, R. L. (1966). Bounds for ceratin multiprocessing anomalies. *Bell System Technical Journal 45*, 1563–1581.

Graham, R. L. (1969). Bounds on multiprocessing timing anomalies. *SIAM Journal on Applied Mathematics 17*, 263–269.

Grünert, T. (1998). *Multi-Level Sequence-Dependent Dynamic Lotsizing and Scheduling*. Aachen: Shaker.

Guinet, A. G. P. and M. Solomon (1996). Scheduling hybrid flowshops to minimize maximum tardiness or maximum completion time. *International Journal of Production Research 34*(6), 1643–1654.

Gupta, J. N. D., K. Krüger, V. Lauff, F. Werner, and Y. N. Sotskov (2002). Heuristics for hybrid flow shops with controllable processing times and assignable due dates. *Computers & Operations Research 29*(10), 1417–1439.

Haase, K. (1994). *Lotsizing and Scheduling for Production Planning*. Berlin: Springer.

Haase, K. (1996). Capacitated lot-sizing with sequence dependent setup costs. *OR Spektrum 18*, 51–59.

Haase, K. (1998). Capacitated lot-sizing with linked production quantities of adjacent periods. In A. Drexl and A. Kimms (eds.), *Beyond Manufacturing Resource Planning (MRP II) – Advanced Models and Methods for Production Planning*, pp. 127–146. Berlin: Springer.

Haase, K. and A. Kimms (2000). Lot sizing and scheduling with sequence dependent setup costs and times and efficient rescheduling opportunities. *International Journal of Production Economics 66*, 159–169.

Harjunkoski, I. and I. E. Grossmann (2002). Decomposition techniques for multistage scheduling problems using mixed-integer and constraint programming methods. *Computers & Chemical Engineering 26*(11), 1533–1552.

Harris, F. W. (1913). How many parts to make at once. *Factory, The Magazine of Management 10*(2), 135–136, 152. Reprinted in *Operations Research (1990) 38*(6), 947–950.

Hax, A. C. and D. Candea (1984). *Production and Inventory Management*. Englewood Cliffs, NJ: Prentice Hall.

Heizer, J. and B. Render (2004). *Principles of Operations Management*. Upper Saddle River, NJ: Prentice Hall.

Helber, S. (1994). *Kapazitätsorientierte Losgrößenplanung in PPS-Systemen*. Stuttgart: M&P Verlag für Wissenschaft und Forschung (in German).

Heuts, R. M. J., H. P. Seidel, and W. J. Selen (1992). A comparison of two lot sizing-sequencing heuristics for the process industry. *European Journal of Operational Research 59*, 413–424.

Hindi, K. S. (1995). Algorithms for capacitated, multi-item lot-sizing without set-ups. *Journal of the Operational Research Society 46*, 465–472.

Holland, J. H. (1975). *Adaptation in Natural and Artificial Systems.* Ann Arbour, MI: The University of Michigan Press.

Hooker, J. N. (1995). Testing heuristics: We have it all wrong. *Journal of Heuristics 1*(1), 33–42.

Hopper, E. and B. Turton (2001). An empirical investigation of meta-heuristic and heuristic algorithms for a 2D packing problem. *European Journal of Operational Research 128*(1), 34–57.

Hurink, J., B. Jurisch, and M. Thole (1994). Tabu search for the job-shop scheduling problem with multi-purpose machines. *OR Spektrum 15*, 205–215.

ILOG (2001). *ILOG OPL Studio 3.5: The Optimization Language.* Paris: ILOG S.A.

Jakobs, S. (1996). On genetic algorithms for the packing of polygons. *European Journal of Operational Research 88*, 165–181.

Jin, Z. H., K. Ohno, T. Ito, and S. E. Elmaghraby (2002). Scheduling hybrid flowshops in printed circuit board assembly lines. *Production and Operations Management 11*(2), 216–230.

Johnson, S. M. (1954). Optimal two- and three-stage production schedules with setup times included. *Naval Research Logistics Quarterly 1*, 61–68.

Kang, S., K. Malik, and L. J. Thomas (1999). Lotsizing and scheduling on parallel machines with sequence-dependent setup costs. *Management Science 45*(2), 273–289.

Katok, E., H. S. Lewis, and T. P. Harrison (1998). Lot sizing in general assembly systems with setup costs, setup times, and multiple constrained resources. *Management Science 44*(6), 859–877.

Kis, T. and E. Pesch (2002). A review of exact solution methods for the non-preemptive multiprocessor flowshop problem. Working Paper, Computer and Automation Research Institute, Hungarian Academy of Sciences, Budapest, Hungary. To appear in *European Journal of Operational Research.*

Kochhar, S. and R. J. Morris (1987). Heuristic methods for flexible flow line scheduling. *Journal of Manufacturing Systems 6*(4), 299–314.

Koulamas, C. and G. Kyparisis (2000). Asymptotically optimal linear time algorithms for two-stage and three-stage flexible flow shops. *Naval Research Logistics 47*(3), 259–268.

Kuhn, H. and D. Quadt (2002). Lot sizing and scheduling in semiconductor assembly – a hierarchical planning approach. In G. T. Mackulak, J. W. Fowler, and A. Schömig (eds.), *Proceedings of the International Conference on Modeling and Analysis of Semiconductor Manufacturing (MASM 2002)*, pp. 211–216. Arizona State University, Tempe, USA.

Kuik, R., M. Salomon, and L. N. Van Wassenhove (1994). Batching decisions: Structure and models. *European Journal of Operational Research* 75(2), 243–263.

Kurz, M. E. and R. G. Askin (2001). An adaptable problem-space-based search method for flexible flow line scheduling. *IIE Transactions* 33(8), 691–693.

Kurz, M. E. and R. G. Askin (2003a). Comparing scheduling rules for flexible flow lines. *International Journal of Production Economics* 85(3), 371–388.

Kurz, M. E. and R. G. Askin (2003b). Scheduling flexible flow lines with sequence-dependent setup times. Working Paper, Clemson University, Clemson, SC. To appear in *European Journal of Operational Research*.

Laguna, M. (1999). A heuristic for production scheduling and inventory control in the presence of sequence-dependent setup-times. *IIE Transactions 31*, 125–134.

Lambrecht, M. R. and H. Vanderveken (1979). Heuristic procedure for the single operation multi-item loading problem. *AIIE Transactions 11*(4), 319–326.

Lee, I., R. Sikora, and M. J. Shaw (1997). A genetic algorithm-based approach to flexible flow-line scheduling with variable lot sizes. *IEEE Transactions on Systems, Man, and Cybernetics – Part B: Cybernetics 27*(1), 36–54.

Leon, V. J. and B. Ramamoorthy (1997). An adaptable problem-space-based search method for flexible flow line scheduling. *IIE Transactions 29*, 115–125.

Lodi, A., S. Martello, and M. Monaci (2002). Two-dimensional packing problems: A survey. *European Journal of Operational Research 141*(2), 241–252.

Maes, J., J. O. McClain, and L. N. Van Wassenhove (1991). Multilevel capacitated lotsizing complexity and LP-based heuristics. *European Journal of Operational Research 53*, 131–148.

Maes, J. and L. N. Van Wassenhove (1986a). Multi item single level capacitated dynamic lotsizing heuristics: A computational comparison (Part I: Static case). *IIE Transactions 18*(2), 114–123.

Maes, J. and L. N. Van Wassenhove (1986b). Multi item single level capacitated dynamic lotsizing heuristics: A computational comparison (Part II: Rolling horizon). *IIE Transactions 18*(2), 124–129.

Maes, J. and L. N. Van Wassenhove (1988). Multi-item single-level capacitated dynamic lot-sizing heuristics: A general review. *Journal of the Operational Research Society 39*(11), 991–1004.

Meyr, H. (1999). *Simultane Losgrößen- und Reihenfolgeplanung für kontinuierliche Produktionslinien – Modelle und Methoden im Rahmen des Supply Chain Management*. Wiesbaden: Gabler (in German).

Meyr, H. (2000). Simultaneous lotsizing and scheduling by combining local search with dual reoptimization. *European Journal of Operational Research 120*(2), 311–326.

Meyr, H. (2002). Simultaneous lotsizing and scheduling on parallel machines. *European Journal of Operational Research 139*(2), 277–292.

Michalewicz, Z. and D. B. Fogel (2000). *How to Solve It: Modern Heuristics.* Berlin: Springer.

Millar, H. H. and M. Yang (1993). An application of lagrangean decomposition to the capacitated multi-item lot sizing problem. *Computers & Operations Research 20*(4), 409–420.

Millar, H. H. and M. Yang (1994). Lagrangian heuristics for the capacitated multi-item lot-sizing problem with backordering. *International Journal of Production Economics 34*(1), 1–15.

Mokotoff, E. (2001). Parallel machine scheduling problems: A survey. *Asia-Pacific Journal of Operational Research 18*(2), 193–242.

Moursli, O. and Y. Pochet (2000). A branch-and-bound algorithm for the hybrid flowshop. *International Journal of Production Economics 64*, 113–125.

Negemann, E. G. (2001). Local search algorithms for the multiprocessor flow shop scheduling problem. *European Journal of Operational Research 128*(1), 147–158.

Néron, E., P. Baptiste, and J. N. Gupta (2001). Solving hybrid flow shop problem using energetic reasoning and global operations. *Omega – The International Journal of Management Science 29*(6), 501–511.

Nowicki, E. and C. Smutnicki (1998). The flow shop with parallel machines: A tabu search approach. *European Journal of Operational Research 106*, 226–253.

Osman, I. H. and G. Laporte (1996). Metaheuristics: A bibliography. *Annals of Operations Research 63*, 513–628.

Özdamar, L. and G. Barbarosoglu (1999). Hybrid heuristics for the multi-stage capacitated lot sizing and loading problem. *Journal of the Operational Research Society 50*, 810–825.

Özdamar, L. and S. I. Birbil (1998). Hybrid heuristics for the capacitated lot sizing and loading problem with setup times and overtime decisions. *European Journal of Operational Research 110*(3), 525–547.

Pinedo, M. (2002). *Scheduling: Theory, Algorithms, and Systems* (2nd ed.). Upper Saddle River, NJ: Prentice Hall.

Pochet, Y. and L. Wolsey (1988). Lot size models with back-logging: Strong reformulations and cutting planes. *Mathematical Programming 40*, 317–335.

Portmann, M.-C., A. Vignier, D. Dardilhac, and D. Dezalay (1998). Branch and bound crossed with GA to solve hybrid flowshops. *European Journal of Operational Research* 107(2), 389–400.

Potts, C. N. and M. Y. Kovalyov (2000). Scheduling with batching: A review. *European Journal of Operational Research* 120, 228–249.

Quadt, D. and H. Kuhn (2003). Production planning in semiconductor assembly. In C. T. Papadopoulos (ed.), *Proceedings of the Fourth Aegean International Conference on Analysis of Manufacturing Systems*, pp. 181–189. University of the Aegean, Samos Island, Greece.

Quadt, D. and H. Kuhn (2004). A conceptual framework for lot-sizing and scheduling of flexible flow lines. Working Paper, Ingolstadt School of Management, Catholic University of Eichstätt-Ingolstadt, Ingolstadt, Germany, submitted.

Rajendran, C. and D. Chaudhuri (1992a). A multi-stage parallel-processor flowshop problem with minimum flowtime. *European Journal of Operational Research* 57(1), 111–122.

Rajendran, C. and D. Chaudhuri (1992b). Scheduling in n-job, m-stage flowshop with parallel processors to minimize makespan. *International Journal of Production Economics* 27, 137–143.

Reeves, C. R. (1995). *Modern Heuristic Techniques for Combinatorial Problems*. London: McGraw-Hill.

Reeves, C. R. and J. E. Rowe (2003). *Genetic Algorithms: Principles and Perspectives: A Guide to GA Theory*. Boston, MA: Kluwer.

Riane, F. (1998). *Scheduling Hybrid Flowshops: Algorithms and Applications*. Ph.D. Thesis, Facultés Universitaires Catholiques de Mons.

Riane, F., A. Artiba, and S. Iassinovski (2001). An integrated production planning and scheduling system for hybrid flowshop organizations. *International Journal of Production Economics* 74(1), 33–48.

Salvador, M. S. (1973). A solution to a special class of flow shop scheduling problems. In S. E. Elmaghraby (ed.), *Symposium on the Theory of Scheduling and Its Applications*, pp. 83–91. Berlin: Springer.

Sawik, T. (1993). A scheduling algorithm for flexible flow lines with limited intermediate buffers. *Applied Stochastic Models and Data Analysis* 9, 127–138.

Sawik, T. (1995). Scheduling flexible flow lines with no in-process buffers. *International Journal of Production Research* 33(5), 1357–1367.

Sawik, T. (2000). Simultaneous versus sequential loading and scheduling of flexible assembly systems. *International Journal of Production Research* 38(14), 3267–3282.

Sawik, T. (2002). Balancing and scheduling of surface mount technology lines. *International Journal of Production Research* 40(9), 1973–1991.

Selen, W. J. and R. M. J. Heuts (1990). Operational production planning in a chemical manufacturing environment. *European Journal of Operational Research 45*, 38–46.

Sivakumar, A. I. and C. S. Chong (2001). A simulation based analysis of cycle time distribution, and throughput in semiconductor backend manufacturing. *Computers in Industry 45*(1), 59–78.

Smith-Daniels, V. L. and L. P. Ritzman (1988). A model for lot-sizing and sequencing in process industries. *International Journal of Production Research 26*, 647–674.

Smith-Daniels, V. L. and D. E. Smith-Daniels (1986). A mixed integer programming model for lot sizing and sequencing packaging lines in the process industries. *IIE Transactions 18*, 278–285.

Soewandi, H. and S. E. Elmaghraby (2001). Sequencing three-stage flexible flowshops with identical machines to minimize makespan. *IIE Transactions 33*(11), 985–993.

Sox, C. R. and Y. Gao (1999). The capacitated lot sizing problem with setup carry-over. *IIE Transactions 31*, 173–181.

Stadtler, H. (1996). Mixed integer programming model formulations for dynamic multi-item multi-level capacitated lotsizing. *European Journal of Operational Research 94*(3), 561–581.

Stadtler, H. (2003). Multilevel lot sizing with setup times and multiple constrained resources: Internally rolling schedules with lot-sizing windows. *Operations Research 51*(3), 487–502.

Sürie, C. and H. Stadtler (2003). The capacitated lot-sizing problem with linked lot-sizes. *Management Science 49*(8), 1039–1054.

Talbi, E.-G. (2002). A taxonomy of hybrid metaheuristics. *Journal of Heuristics 8*(5), 541–564.

Tempelmeier, H. (1997). Resource-constrained materials requirements planning – MRP rc. *Production Planning & Control 8*(5), 451–461.

Tempelmeier, H. (2003). *Material-Logistik – Modelle und Algorithmen für die Produktionsplanung und -steuerung und das Supply Chain Management* (5th ed.). Berlin: Springer (in German).

Tempelmeier, H. and M. Derstroff (1996). A lagrangean-based heuristic for dynamic multilevel multiitem constrained lotsizing with setup times. *Management Science 42*(5), 738–757.

Thizy, J.-M. and L. N. Van Wassenhove (1985). Lagrangean relaxation for the multi-item capacitated lot-sizing problem: A heuristic implementation. *IIE Transactions 17*(4), 308–313.

Thonemann, U. W. and J. R. Bradley (2002). The effect of product variety on supply-chain performance. *European Journal of Operational Research 143*, 548–569.

Trigeiro, W. W., L. J. Thomas, and J. O. McClain (1989). Capacitated lot sizing with setup times. *Management Science 35*(3), 353–366.

Tsubone, H., M. Suzuki, T. Uetake, and M. Ohba (2000). A comparison between basic cyclic scheduling and variable cyclic scheduling in a two-stage hybrid flow shop. *Decision Sciences 31*(1), 197–222.

Verma, S. and M. Dessouky (1999). Multistage hybrid flowshop scheduling with identical jobs and uniform parallel machines. *Journal of Scheduling 2*(3), 135–150.

Wardono, B. and Y. Fathi (2003a). A tabu search algorithm for the multistage parallel machine problem with limited buffer capacities. Working Paper, North Carolina State University, Raleigh, NC. To appear in *European Journal of Operational Research*.

Wardono, B. and Y. Fathi (2003b). A tabu search algorithm for the multistage parallel machine problem with unlimited buffer capacities. Working Paper, North Carolina State University, Raleigh, NC. To appear in *European Journal of Operational Research*.

Wittrock, R. J. (1985). Scheduling algorithms for flexible flow lines. *IBM Journal of Research and Development 29*(4), 401–412.

Wittrock, R. J. (1988). An adaptable scheduling algorithm for flexible flow lines. *Operations Research 36*(4), 445–453.

Yang, T., H.-P. Fu, and C. Yang (2002). A simulation-based dynamic operator assignment strategy considering machine interference – a case study on integrated circuit chip moulding operations. *Production Planning & Control 13*(6), 541–551.

Acknowledgment

This research was supported by grant 03KUM1EI of the Federal Ministry for Education and Research (Bundesministerium für Bildung und Forschung, BMBF), Germany.

Lecture Notes in Economics and Mathematical Systems

For information about Vols. 1–454
please contact your bookseller or Springer-Verlag